Y0-ARU-926

HISTORICAL BIOGEOGRAPHY

JORGE V. CRISCI
LILIANA KATINAS
PAULA POSADAS

HISTORICAL BIOGEOGRAPHY

AN INTRODUCTION

HARVARD UNIVERSITY PRESS
Cambridge, Massachusetts London, England 2003

Printed in the United States of America

Library of Congress Cataloging-in-Publication Data

Crisci, Jorge Victor.
Historical biogeography : an introduction / Jorge V. Crisci, Liliana Katinas, Paula Posadas.
p. cm.
Includes bibliographical references and index.
ISBN 0-674-01059-0 (cloth : alk. paper)
1. Biogeography—History. I. Katinas, Liliana. II. Posadas, Paula. III. Title.

QH84 .C6798 2003
578′.09—dc21 2002192236

PREFACE

In a 1959 essay entitled "The Voice of Poetry in the Conversation of Mankind," Michael Oakeshott develops the notion of knowledge as a community-owned social construct that is the result of our ability to participate in an unending conversation. Oakeshott says: "As civilized human beings, we are the inheritors, neither of an inquiry about ourselves and the world, nor of an accumulating body of information, but of a conversation, begun in the primeval forests and made more articulate in the course of centuries. It is a conversation which goes on both in public and within each of ourselves."

According to Oakeshott, education, properly speaking, is an initiation into the skills and partnership of this conversation in which we learn to recognize the voices, to distinguish the proper occasions of utterance, and in which we acquire the intellectual and moral habits appropriate to conversation. And it is this conversation that, in the end, characterizes every human activity and utterance.

Each voice reflects a human activity, begun without premonition of where it would lead, but acquiring for itself in the course of the engagement a specific personality and manner of speaking. Over time, each voice modulates in reaction to those around it.

Among the voices of biology, historical biogeography recently has acquired, or begun to acquire, an authentic voice and language of its own. Our purpose is to consider the voice of historical biogeography: its utterances, manners of speaking, modulation, and manner of thinking, which in this book are influenced strongly by histories of South America.

To listen to the voice of historical biogeography using empirical examples from South America is to return to the birth of evolutionary theory. Darwin himself stated in the opening paragraph of *The Origin of the Species* (1859): "When on board H.M.S. *Beagle,* as naturalist, I was much struck with certain facts in the distribution of the inhabitants of South America, and in the geological relations of the present to the past inhabitants of that continent. These facts seemed to me to throw some light on the origin of species."

Today, as in Darwin's time, the distribution of living beings offers an inexhaustible source of light on the evolution of life on Earth. There are few facets of evolutionary biology that cannot be illuminated by the study of the history of these distributions, otherwise known as historical biogeography. Furthermore, historical biogeography is passing through an extraordinary revolution encompassing its fundamental principles, basic concepts, methods, and relationships with other disciplines of comparative biology.

In this book we explain and illustrate the fundamentals and the most frequently used methods of historical biogeography, including how to recognize when one has a research problem that requires a historical biogeographic approach; how to decide upon the most appropriate kind of data to collect; how to choose the best method for the problem at hand; how to perform the necessary calculations, and if a computer program is needed, which one to use; and how to interpret the results. It is not our goal to suggest the adoption of a single method, but to elucidate the biological assumptions of each method.

We include case studies, selected mainly from our own research.

These studies encompass a variety of research goals and contexts and give an overall impression of how these methodologies are used.

Although this book is primarily a text for researchers and students of biology, it may also interest those in such fields as geology and geography, since the voice of historical biogeography echoes in many sciences.

We would like to thank A. Bartoli, M. Bonifacino, M. Donato, M. Ebach, M. Heads, P. Hoch, P. Ladiges, D. Miranda-Esquivel, G. Nelson, F. Ocampo, E. Ortiz Jaureguizar, J. Patton, S. Roig-Juñent, R. Tortosa, and G. Voelker for comments on the manuscript or parts of the manuscript. We greatly appreciate Piero Marchionni and Mariano Donato's help in the preparation of this book. We would like to acknowledge the invaluable help of Lucy Gómez de Mainer who spent a lot of her time improving our English. Hugo Calvetti prepared the illustrations for this book. We would also like to thank our editor, Michael Fisher, and our manuscript editor, Kate Brick, for encouragement and professional assistance. The advice of Brian Farrell, Gary Nelson, Jim Wilgenbusch, and one anonymous reviewer have undoubtedly made this book of much higher quality that it would have been otherwise. Those faults that still remain are entirely our responsibility. Our research on biogeography was supported by National Geographic Society (Grants #3966–88, 4662–91, 5776–96); Consejo Nacional de Investigaciones Científicas y Técnicas (CONICET), Argentina; and Agencia Nacional de Promoción Científica y Tecnológica (PICT99 N 6866). Permission to reprint an excerpt by Jorge Luis Borges from "Avatars of the Tortoise," in *Labyrinths*, copyright © 1962, 1964, has been granted by New Directions Publishing Corporation. Finally, Victoria Crisci, Elena Katinas, and Edgardo Ortiz Jaureguizar have assisted us and supported us throughout this project. We may never be able to repay them for all their help, encouragement, and extraordinary patience.

CONTENTS

We (the undivided divinity operating within us) have dreamt the world. We have dreamt it as firm, mysterious, visible, ubiquitous in space and durable in time; but in its architecture we have allowed tenuous and eternal crevices of unreason which tell us it is false.

Jorge Luis Borges, *Avatars of the Tortoise*

INTRODUCTION: WHAT IS HISTORICAL BIOGEOGRAPHY?

INVESTIGATIVE STUDIES are often characterized by a central plot or metaphor that provides solid ground in which theories root themselves (Haraway, 1976). Such metaphors serve to bridge abstractions and the real world (Hesse, 1966). In the last decade, a metaphor created by the Italian-French botanist Léon Croizat (1964) has unified the field of comparative biology. His metaphor views biological diversity as a historical fact that occurs in three dimensions: form, space, and time.

Thus, biological diversity is understood as a result of the history of life upon Earth expressed through its changes of form in space and time. The term "form" refers not only to the morphological characters of living beings, but also to others, such as molecular ones (DNA sequences, for example). Systematics is the part of comparative biology that stresses the form, paleontology and embryology stress the time, and biogeography stresses the space.

DEFINITION AND CONCEPTS

Biogeography may be simple to define—the study of the geographic distribution of living beings—but this apparent simplicity hides a great complexity. Biogeography goes further than the classic disciplines to include such subjects as geology, geography, and biology. Thus, it is not surprising that biogeography means different things to different researchers.

For convenience, biogeographers have recognized two traditions in biogeographic investigation, ecological biogeography and historical biogeography. Swiss botanist Agustin P. de Candolle (1820) was the first to distinguish these two traditions. According to his definition, the explanations for ecological biogeography depend upon physical causes that are acting in the present time, whereas the explanations for historical biogeography depend upon causes that existed in the past.

Thus, ecological biogeography studies how ecological processes that happen in short periods of time act on the distributional patterns of living beings, whereas historical biogeography studies how those processes that happen over long periods of time—through million of years—(for example, evolution or tectonics) influence known patterns (Cox & Moore, 1993). Some authors place the study of the biogeographic effects of Pleistocene glaciations between ecological and historical biogeography (Myers & Giller, 1988).

Theories, hypotheses, and models have been postulated in each one of these two traditions, but unfortunately with little interaction between them. This lack of communication reflects the past predominance of narrative over analytical biogeographic methods. The narrative method allows authors to base their conclusions on beliefs more than on rigorous inferences. When analytical methods are applied in biogeography we find that the organism's distributional patterns are not the result of a single cause, be it ecological or historical. The present division between ecological and historical biogeography is occasional and it is very possible that they may be joined in a research program in the future. We expect that the approach to historical biogeography presented here will be helpful in advancing that much-needed synthesis.

External and Internal Forces within the Discipline

Numerous forces are shaping this rapidly evolving discipline (Crisci, 2001). The external developments include global tectonics as the dominant paradigm in geosciences, phylogeny as the basic language of comparative biology, molecular systematics as a new window onto nature, and the biologist's perception of biogeography.

Geographic stasis was the big question during most of the twentieth century. This was a question inherited from the nineteenth century and the activities of the early biogeographers (Nelson & Platnick, 1984). Alfred Wegener in 1915 first proposed the idea of continental movement. Because the specific mechanism Wegener proposed to account for continental movement was not feasible, his theory fell into disfavor with the great majority of geologists. In the early 1960s, new evidence developed that provided a mechanism for continental movement and crustal evolution, from which sprang the rejuvenated field of geoscience (Condie, 1997). The biogeographic consequences of plate movements and interactions are enormous. The rearrangement of continental land masses and islands and the opening and closing of sea and ocean basins initiated by these movements and interactions have profoundly affected the distribution and history of organisms. Therefore, the whole idea of Earth evolution has a strong influence on biogeography, reflected in the motto of Léon Croizat (1964): "Earth and life evolve together."

The next of these external forces, the study of phylogeny, resulted in the cladograms used today in comparative biology (Nelson & Platnick, 1981; Swofford et al., 1996; Kitching et al., 1998; Schuh, 2000). Cladograms are a powerful method of communicating a system of relationships to other biologists (Morrone et al., 1992; Crisci, 1992, 1998a; Crisci & Katinas, 1997; Katinas & Crisci, 1999). Biologically and historically, the phylogenetic relationships between taxa and their geographic distribution are intimately linked. Nodes of a cladogram are potentially informative about the distributional history of the organisms and about relationships among geographic areas occupied by them (Crisci, 1998b). For this reason, phylogenetic inference plays a crucial role in historical biogeography. On the other hand, the increased use of quantifiable phylogenetic methods and statistical hypothesis-testing is forcing biogeographers toward a more precise formulation of methodological practices and theoretical ideas and the exact quantification of their implications.

While methods for phylogenetic estimation were developing in the 1960s, another revolution was happening in molecular biology. Methods for examining the molecular structure of proteins and nucleic acids were soon adopted by evolutionary biologists, and the data available for phylogenetic estimation began to increase exponentially (Hillis et al., 1996a). Molecular methods have provided alternatives to morphological data in phylogeny reconstruction. Large number of individuals can be sampled in a relatively short period of time, and each of these samples may be examined for potentially thousands of discrete characters at the restriction site or at nucleotide level. These characters are largely uncoupled from environmental or developmental influences, and may be compared to distantly related taxa. Phylogenetic reconstructions based on molecular datasets provide opportunities for the study of evolutionary phenomena. We discuss the role of molecular data in historical biogeographic studies in greater detail in chapter 12.

Finally, biogeography (as a whole, not only historical biogeography) is perceived as an oddity by a vast majority of biologists (Crisci et al., 2000), and has been influenced by their opinions of it. Its extraordinary complexity and its diversity of approaches make biogeography an unusual offshoot of biology. A quotation from Gareth Nelson (1978) reflects this perception: "Biogeography is a strange discipline. In general, there are no institutes of biogeography; there are no departments of it. There are no professional biogeographers—no professors of it, no curators of it. It seems to have few traditions. It seems to have few authoritative spokesmen."

Internally, the forces that are shaping historical biogeography include the proliferation of competing articulations (for example, ecology versus history; panbiogeography versus cladistic biogeography; event-based methods versus pattern-based methods) (Crisci & Morrone, 1992a), and recourse to philosophy and the debate over fundamentals (for example, conceptions of space—absolute space versus relative space).

A revolution in science can be recognized by old terms acquiring new meanings and by an increase in philosophizing by its practitioners (Heisenberg, 1958). Historical biogeography is clearly in the midst of a revolution and this is nowhere more evident than in the fact that, of the thirty-one techniques of historical biogeography currently in use, twenty-four (77 percent) have been proposed in the last fourteen years. Furthermore, in the last years of the twentieth century three books articulating different points of view on the subject were published: *Panbiogeography: Tracking the History of Life* (Craw et al., 1999); *Cladistic Biogeography* (Humphries & Parenti, 1999); and *Phylogeography: The History and Formation of Species* (Avise, 2000).

This revolution may well testify to the health of the subject, but more promisingly, it may presage major advances in the field, as Thomas Kuhn's (1970) theory may predict: "The proliferation of competing articulations, the willingness to try anything, the expression of explicit discontent, the recourse to philosophy and to debate over fundamentals, all these are symptoms of a transition from normal to extraordinary research." We now move on from a discussion of the origins and history of biogeography to a discussion of its scope and some of its components.

SPATIAL ANALYSIS

Spatial analysis is simply the study of phenomena that manifest themselves in space. It deals with formal models of spatial organization (Gatrell, 1983), such as spatial arrangement, space-time processes, and spatial prediction or spatial postdiction. By spatial arrangement is meant the pattern and location of the objects under study. The study of time-space processes concerns how spatial arrangements are modified by movement or spatial interaction. Spatial prediction and postdiction seek respectively to forecast future spatial arrangements or to establish on the basis of present evidence what the past spatial arrangements must have been like.

This threefold scheme can be applied to historical biogeography: Spatial arrangement is the distribution of organisms throughout geographic space, space-time processes are the events that can modify the geographic arrangement of the organisms, and postdiction is the possibility of determining past biogeographic events in terms of given observations (prediction, of course, is not very useful in historical biogeography, but currently is of great interest in ecological biogeography—see for example Sala et al., 2000).

Spatial Arrangement

No one will deny the value of understanding the geographic distribution of organisms in biogeography, but unfortunately most books in the field approach the subject from an ecological perspective, emphasizing ecosystem rather than areas (for an example, see Cox & Moore, 1993). Recently, there has been an attempt (Craw et al., 1999) to reintroduce and reemphasize the importance of the spatial or geographic dimension of life's diversity for biogeography and for our understanding of evolutionary patterns and processes. Despite its partisan adoption of a particular method, this attempt, based on an approach called panbiogeography (see chapter 5), has two important assets for biogeography: It focuses on the role of locality and place in the history of life, and it considers that an understanding of locality is a fundamental precondition to any adequate analysis of historical biogeography.

Space-Time Processes

Biogeographers have identified three different space-time processes that can modify the geographic spatial arrangement of organisms: extinction, dispersal, and vicariance. Biogeographers have accepted without controversy the influence of extinction—the death of all individuals in a local population, a species, or a higher taxon—on the geographic distribution of organisms. This is not the case for the other two processes (dispersal and vicariance). These have been considered for many years to be competing explanations of the geographic distribution of organisms. Every time a biogeographer tries to explain disjunct distribution patterns such as the one shown by the plant genus *Nothofagus* (southern beeches, Nothofagaceae—see chapter 11), there are disagreements about the origin of the distribution. Either its common ancestor originally occurred in one of the areas and later dispersed into the other ones, where descendants survive to the present day, or its ancestor was originally widespread in greater areas, which became fragmented, and its descendants have survived in the fragments until now. These historical explanations are named, respectively, dispersal and vicariance (Nelson & Platnick, 1984).

In the dispersal explanation, the range of the ancestral population was limited by a pre-existing barrier, which was crossed by some of its members. If they colonize the new area and remain isolated from the original population, they may eventually differentiate into a new taxon. In the vicariance explanation, the ancestral population was divided into subpopulations by the development of barriers they cannot cross. The appearance of the barrier causes the disjunction, so the barrier cannot be older than the disjunction. In the dispersal explanation the barrier is older than the disjunction (Fig. I.1).

For centuries dispersal was the dominant explanation for the distribution of organisms, based on strict adherence to the geological concept of Earth stability. Two botanists, Stanley Cain (1944) and Léon Croizat (1958), in particular, were among the first scientists to challenge vocally the dispersal explanation as the main process in biogeography and promote vicariance as an equally important process. Currently both vicariance and dispersal are recognized as significant biogeographic processes, but neither takes precedence over the analysis of distributional patterns.

Recently, Fredrik Ronquist (1997b) has suggested the need to separate dispersal into two kinds of events. According to him, dispersal in response to the disappearance of a previous dispersal barrier (predicted dispersal = range expansion) should be separated from random colonization of disjunct areas (random dispersal). Bruce S. Lieberman (2000) has proposed the term "geodispersal" to refer to a particular kind of dispersal event that results in congruent patterns among different lineages. Geodispersal does not imply dispersal over a barrier, but refers to episodes of range expansion occurring simultaneously in different clades, and implicates a close association between patterns of range expansion and geological events. Lieberman defines geodispersal as "the expansion of the range of a group of species due to the elimination

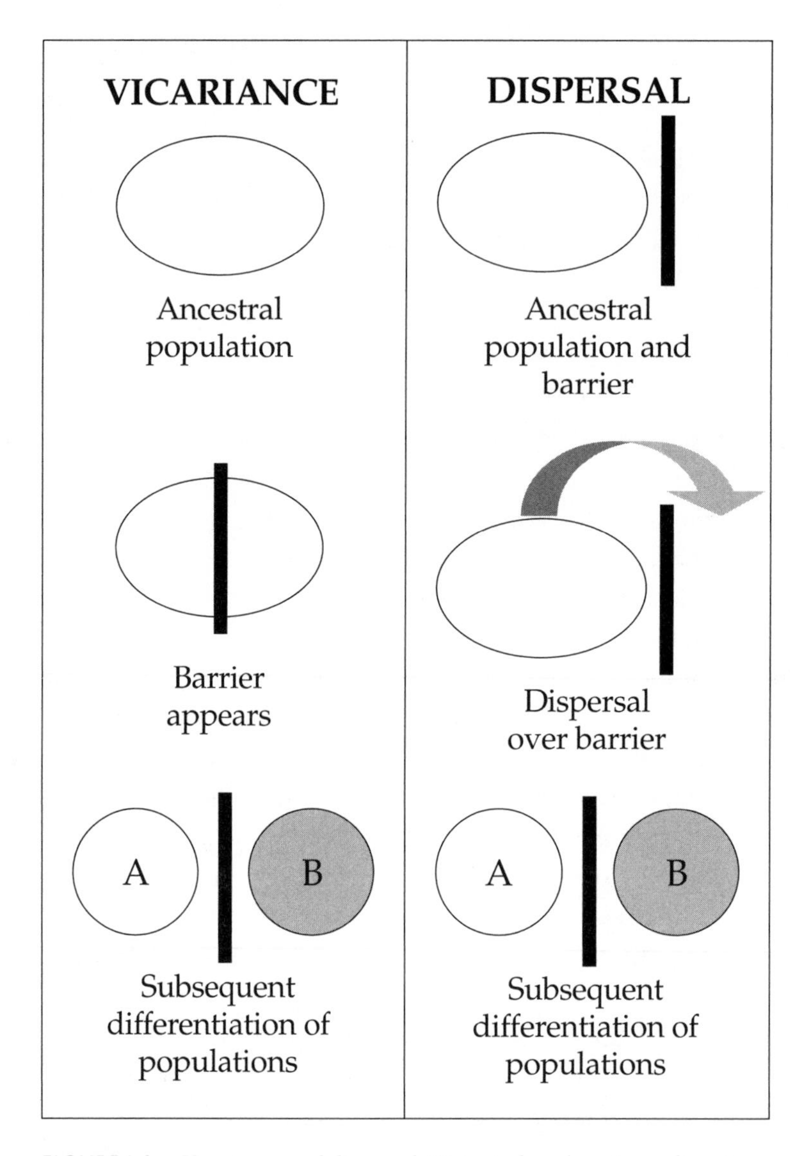

FIGURE I.1. Vicariance and dispersal. Historical explanations of two-taxa disjunct distribution.

of some topographic barrier followed by the emergence of a new barrier which produces subsequent vicariance" (Lieberman 2000). The geodispersal concept is closely associated with Ronquist's predicted dispersal.

The distinction among different kinds of dispersal by Ronquist and

Lieberman is not taken into account by current historical biogeographic approaches (except by the technique proposed by Ronquist [1997a] called constrained DIVA—see chapter 8), but they will play a role in future methodological developments, especially in event-based methods (see chapter 8) in which the different events are weighted by a cost assignment.

Vicariance can be subdivided into three kinds of events: vicariance followed by speciation (this process leads to sister species distributed in sister areas); vicariance events that lack allopatric speciation (this process retains widespread species); and vicariance events followed by speciation after a previous speciation event independent of the vicariance of the area. According to Gareth Nelson and Pauline Ladiges (1996), this latter process leads to geographic paralogy, multiplication of lineages, sympatry, and redundancy (Fig. I.2).

In conclusion, it is interesting to point out that according to Croizat (1964), the vicariance-dispersal opposition can be resolved by applying a biogeographic model involving alternating cycles of dispersal and vicariance (vicariance form-making or periodic mobilism). This model was empirically exemplified by R. C. Craw and colleagues (1999:17) through the admiral butterfly genera *Bassaris* and *Vanessa.*

Spatial Postdiction

Reconstructing past biogeographic events in historical biogeography can be accomplished from three different angles: reconstructing the distributional history of individual groups ("taxon biogeography"), reconstructing the history of areas of endemism (the search for general area relationships or "area biogeography"), and reconstructing the distributional history of biotas (search for spatial homology). Unfortunately, these different objectives are often confused in historical biogeographic applications. It is not unusual to find an intention to reconstruct the biogeographic history of a taxon hidden as the reconstruction of the history of the areas of endemism. In the next section we will try, among other objectives, to remedy this situation by clarifying the different historical biogeographic approaches.

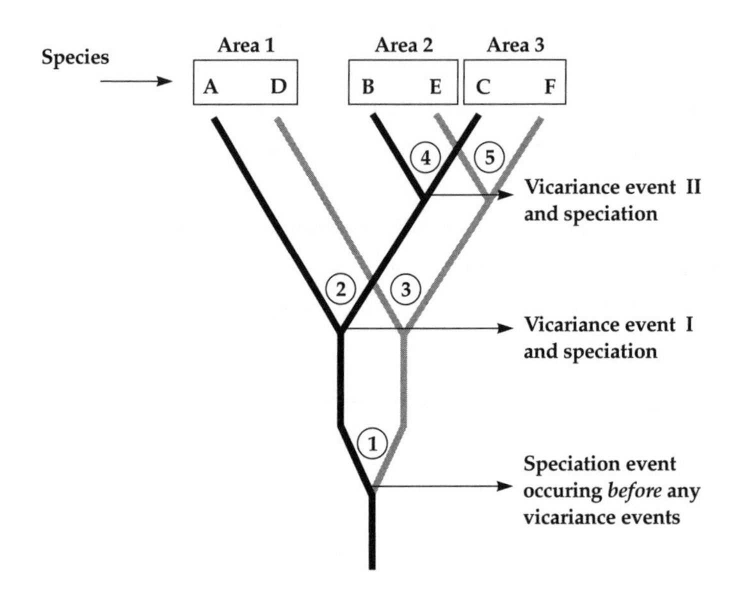

FIGURE I.2. Geographic paralogy as a result of a speciation event (1) independent of area vicariance. Two subsequent speciation events (2, 3) are related with vicariant event I, and two subsequent speciation events (4, 5) are related with vicariant event II. Species: A, B, C, D, E, F. Areas: 1, 2, 3.

TAXONOMY OF METHODS

The proliferation of competing articulations in historical biogeography has generated a great number of approaches to the subject. This diversity is difficult to present without some kind of taxonomy of methods. The taxonomy that we present, as all taxonomies, is debatable, but it is used here as a way to organize the prevailing tangled state of the discipline. Other taxonomies of methods have been proposed by Andersson (1996),

Humphries (2000), Lieberman (2000), van Veller and colleagues (2000), and Ebach and Edgecombe (2001).

In chapters 2–10 we develop nine basic historical biogeographic approaches (Crisci, 2001):

1. *Center of origin and dispersal.* This approach originated in the Darwin-Wallace tradition. They considered that species originate in one center of origin, from which some individuals subsequently disperse by chance, and then change through natural selection. Among its most prominent exponents was William D. Matthew (1915).
2. *Phylogenetic biogeography.* This is also a dispersalist approach and the first one to consider a cladogram for a given group of organisms as the basis for inferring its biogeographic history (Brundin, 1966). Phylogenetic biogeography can be defined as the study of the history of monophyletic groups in time and space.
3. *Ancestral areas.* This approach was formalized by Kåre Bremer (1992) and also uses cladograms as raw data in a dispersalist view. The procedure allows one to identify the ancestral area of a group from the topological information of its cladogram given the information on their presence on deep and numerous branches in that cladogram. Each area can be considered a binary character with two states (present or absent) and optimized onto the cladogram. By comparing the numbers of gains and losses, it is possible to estimate the areas most likely to have been part of the ancestral areas.
4. *Panbiogeography.* This approach, originally proposed by Croizat (1958), plots distributions of organisms on maps and connects together the disjunct distribution areas or collection localities with lines called tracks. Individual tracks for unrelated groups of organisms are then superimposed, and if they coincide, the resulting summary lines are considered generalized tracks. Generalized tracks indicate the preexistence of ancestral biotas, which subse-

quently become fragmented by tectonic and/or climate changes. The area where two or more generalized tracks intersect is called a node. It means that different ancestral biotic and geological fragments interrelate in space/time, as a consequence of terrane collision, docking, or suturing, thus constituting a composite area.

5. *Cladistic biogeography.* This approach was originally developed by Donn Rosen (1978) and Gareth Nelson and Norman Platnick (1981) and it considers both vicariance and dispersal. Cladistic biogeography assumes that the correspondence between phylogenetic relationships and area relationships is biogeographically informative. Comparisons between area cladograms derived from different taxa that occur in a certain region allow general patterns to be elucidated. A cladistic biogeographic analysis comprises two steps: The construction of area cladograms from different taxon cladograms, and the derivation of a general area cladogram or cladograms.
6. *Parsimony analysis of endemicity (PAE).* This approach is a tool of historical biogeography that helps to make clear the patterns of organism distributions using biota similarity (B. Rosen, 1988). The PAE classifies localities, quadrats, or areas (analogous to taxa, if compared with the analysis of phylogenetic systematics) according to their shared taxa (analogous to characters) by means of the most parsimonious solution (parsimony principle). Occurrence of a particular taxon in an area can be interpreted as a character. Shared presences of taxa are treated as synapomorphies in cladistic analysis. This approach, originally proposed by Brian Rosen, does not make assumptions about processes; however, according to Craw (1988a), the "character" reversions in the resulting cladograms could be biogeographically interpreted as extinctions, whereas the parallelisms could be interpreted as dispersals.
7. *Event-based methods.* This approach postulates explicit models of

the processes that have an effect on the geographic distribution of living organisms (Ronquist, 1997b). The different types of processes (dispersal, extinction, and vicariance) are identified and assigned values of benefit and cost under an explicit nature-functioning model. Consequently, the distributional history of a taxon is inferred on the basis of phylogenetic information and applying the criterion of maximum benefits and minimum costs in regard to biogeographic processes (for example, maximizing vicariance and minimizing dispersal and extinction).

8. *Phylogeography.* This approach was originally proposed by John Avise and colleagues (1987) and is the study of the principles and processes governing the geographic distribution of genealogical lineages at the intraspecific level using mitochondrial DNA (mtDNA) in animals and mainly chloroplast DNA (cpDNA) in plants. In this approach the individuals are genotyped and assigned to maternal lineages and the resulting phylogeny is related to patterns of geographic distribution.
9. *Experimental biogeography.* This approach, proposed by Haydon, Radtkey, and Pianka (1994), exploits computers to model faunal build-up repeatedly against a fixed vicariant background over ecological and evolutionary time scales. This approach enables a biogeographer to know both vicariant history and actual phylogeny. Moreover, history can be replayed repeatedly to accumulate a sample of multiple phylogenies and to estimate probability density functions for biogeographic variables. Roles of stochastic, historical, and ecological processes in adaptive radiations can also be assessed.

The nine aforementioned approaches could be described further from many different perspectives. We will next discuss six of these perspectives, summarized in Table I.1, which illustrate the multitude of small differences among the nine approaches.

Table I.1. Main characteristics of the nine historical biogeographic approaches (N/A = not applicable).

	Center of origin and dispersal	Phylogenetic biogeography	Ancestral areas	Panbio-geography	Cladistic biogeography	Parsimony analysis of endemicity	Event-based methods	Phylo-geography	Experimental biogeography
Processes									
Dispersal	✓	✓	✓			N/A			
Dispersal and vicariance				✓	✓	N/A	✓	✓	✓
Extinction	✓	✓	✓	✓	✓	N/A	✓	✓	✓
History									
Area					✓	✓		✓	✓
Taxon	✓	✓	✓		✓		✓	✓	
Biota				✓					
Phylogeny		✓	✓		✓		✓	✓	✓
Explicit model							✓		✓
Rank									
Below species level			✓	✓		✓		✓	✓
Species level or above	✓	✓	✓	✓	✓	✓	✓		✓
Biota similarity						✓			

Processes

What processes do these approaches take into account? Three approaches assume only the possibility of extinction and dispersal: center of origin and dispersal, phylogenetic biogeography, and ancestral areas. Panbiogeography, cladistic biogeography, event-based methods, phylogeography, and experimental biogeography assume the possibility of extinction, dispersal, and vicariance. Only parsimony analysis of endemicity does not assume the possibility of extinction, because it does not make assumptions about processes.

History

What kind of history is it that the approaches try to reconstruct? Panbiogeography's main concern is the history of biotas. Four approaches deal mainly with the distributional history of individual groups: center of origin and dispersal, phylogenetic biogeography, ancestral areas, and event-based methods. Two, parsimony analysis of endemicity and experimental biogeography, deal with the history of areas. The two remaining approaches, cladistic biogeography and phylogeography, deal with the history of individual groups and areas.

Phylogeny

How do these approaches relate to phylogeny? Six of the nine approaches discussed use taxon cladograms as a basic tool of their methodology: phylogenetic biogeography, ancestral areas, cladistic biogeography, event-based methods, phylogeography, and experimental biogeography.

Explicit Models

Which of the approaches use explicit biogeographic models? Two approaches, event-based methods and experimental biogeography, have an explicit model of cost and benefit values for biogeographic events. One of them, experimental biogeography, is a computer simulation method.

Taxonomic Rank

Above or below what taxonomic rank are these approaches applied? Phylogeography is applied exclusively below species level. Four approaches are applied mainly at species level or above: center of origin and dispersal, phylogenetic biogeography, cladistic biogeography, and event-based methods. The other four, ancestral areas, panbiogeography, parsimony analysis of endemicity, and experimental biogeography, are applied at any level.

Biota Similarity

Finally, which of these approaches uses biota similarity as a biogeographic method? Parsimony analysis of endemicity classifies areas by their shared taxa, analyzed according to the most parsimonious solution. Occurrence of a particular taxon in an area can be interpreted as a character. Shared presences of taxa are treated as synapomorphies in cladistic analysis.

The astonishing complexity of historical biogeography can be illustrated by the diversity of techniques employed. In Table I.2 we have classified thirty-one techniques employed in historical biogeography according to which of the aforementioned approaches they fall under. Quite a few (twelve) fall under the cladistic biogeographic approach.

As we mentioned, all taxonomy of historical biogeographic methods is debatable. Thus, for instance, the reconciled trees method included here as an event-based method has been considered elsewhere to be a method of cladistic biogeography. Other techniques whose placement could be considered dubious are two of the ancestral area methods that apply Fitch optimization with or without weighting. These two methods could be included as event-based methods because the application of these optimizations takes place on the hypothesis of a nature-functioning model; the same could be said of the WISARD technique of cladistic biogeography.

Table I.2. Historical biogeographic techniques listed under the corresponding approaches and with their original authors

Techniques	Authors
Center of origin and dispersal	Matthew, 1915
Phylogenetic biogeography	Brundin, 1966
Ancestral areas	
Camin-Sokal optimization	Bremer, 1992
Fitch optimization	Ronquist, 1994
Weighted ancestral areas analysis	Hausdorf, 1998
Panbiogeography	
Track analysis	Croizat, 1958
Spanning graphs	Page, 1987
Track compatibility	Craw, 1988a
Cladistic biogeography	
Reduced area cladogram	D. Rosen, 1978
Ancestral species map	Wiley, 1980
Quantitative phylogenetic biogeography	Mickevich, 1981
Component analysis	Nelson & Platnick, 1981
Brooks parsimony analysis	Wiley, 1987
Component compatibility	Zandee & Roos, 1987
Quantification of component analysis	Humphries et al., 1988
Three-area statement	Nelson & Ladiges, 1991a
Integrative method	Morrone & Crisci, 1995
WISARD	Enghoff, 1996
Paralogy-free subtrees	Nelson & Ladiges, 1996
Vicariance events	Hovenkamp, 1997
Parsimony analysis of endemicity	
Localities	B. Rosen, 1988
Areas of endemism	Craw, 1988a
Quadrats	Morrone, 1994a
Event-based methods	
Coevolutionary two-dimensional cost matrix	Ronquist & Nylin, 1990
Reconciled trees	Page, 1994a,b
Dispersal-vicariance analysis	Ronquist, 1997b
Jungles	Charleston, 1998
Bayesian approach to cospeciation	Huelsenbeck et al., 2000a
Combined method	Posadas & Morrone, 2001
Phylogeography	Avise et al., 1987
Experimental biogeography	Haydon, Radtkey, & Pianka, 1994

Finally, regarding the kind of history the approaches try to reconstruct, Mike Crisp (2001) asks if there is a real difference between the objectives of area biogeography and the search for spatial homology (shared history of whole biotas).

I

METHODS IN HISTORICAL BIOGEOGRAPHY

1

DISTRIBUTION AREAS AND AREAS OF ENDEMISM

THE INITIAL PHASE of biogeographic studies consists of analyzing the geographic distribution areas of certain taxa. Some of the first questions biogeographers pose are: Why do some taxa have a widespread distribution whereas others have a more restricted distribution (endemic taxa)? How can we explain disjunct distributions in which members of the same taxon inhabit localities very far away from one another without any geographic continuity? Why is a taxon richer in species in some regions than in others? Why is the biota of one region more diverse than the biota of other regions?

To answer most of these questions it is necessary to delimit the study areas, and to do so, two different concepts must be considered—distribution areas and areas of endemism. The study area must be defined before a biogeographic method can even be selected, because some methods use distribution areas (for example, some variation of parsimony analysis of endemicity [PAE]), others areas of endemism (for example, cladistic biogeography), and still others do not always require any area delimitation (for example, panbiogeography).

DISTRIBUTION AREAS

The distribution area is the total region within which any taxonomic unit is distributed or presents itself (Cain, 1944). The distribution area of a taxon is related to several factors, such as climate (essentially temperature and humidity), habitat characteristics, and intra- and interspecific competition.

Biogeographers record the distribution area description of a species by transcribing it on a map. The simplest description is a cluster of locality points, which in solely descriptive terms constitutes a sufficient representation of the geographic distribution of a species. Later, the area can be delimited simply by enclosing the points with a line (Morrone et al., 1994). There are also more precise methods, such as the cartographic and areographic methods (Zunino & Zullini, 1995; Zunino, 2000). The cartographic method consists of establishing quadrats (Rapoport, 1975) on a map and filling in the quadrats where the species are (Morrone et al., 1996). Among the areographic methods, that of mean propinquity (Rapoport, 1975; Rapoport & Monjeau, 2001) consists of connecting the neighboring distribution points marked on a map by means of arcs. Then the mean of the distance between the localities (the mean of the arcs) must be established, and then every point is compassed around whose ratio must be equal to the obtained mean. That leads to the formation of colonies of maximum propinquity that must be connected with the nearest colonies. The result is a maximum propinquity tree—a tree with all the nodes connected, but without circuits.

Generally the methods to delimit distribution areas are simplifications of organism distribution in nature, and they do not usually represent the real distribution area. That is because often much of the data used to reconstruct a distribution area comes from literature (for example, revisions, monographs), museum specimens, and biological inventories. These data are transported to a distribution map that consists of a model area (Udvardy, 1969), which will necessarily imply a simplification of reality. This simplification is due to several factors; for instance, in most cases the representations are bidimensional, or the localities chosen represent only a sampling of a taxon distribution area. The distribution maps may be presented simply as points that represent the localities where the taxon has been found, but when it is necessary to compare the distribution of two or more taxa a more detailed approach is usually needed, for example the cartographic and areographic methods.

AREAS OF ENDEMISM

Most historical biogeographic studies use the area of endemism as an analysis unit—a concept that differs from that of distribution area because it implies that the distributions of two or more taxa overlap. Agustin de Candolle (1838) introduced the concept of the area of endemism when in his book about the distribution of Asteraceae he wrote: "These regions were not established *a priori;* I have only recognized as such those areas that are naturally defined and in which I have seen several endemic species."

Delimiting areas of endemism is one of the fundamental steps and one of the most problematical in a historical biogeographic analysis. Barbro Axelius (1991) said that the areas to be used in a cladistic biogeographic analysis must be areas of endemism, and that if this is not accomplished the results of the analysis could be meaningless. Coincidentally, Kåre Bremer (1993) suggested that the delimitation of areas is a methodological problem which deserves much more attention in cladistic biogeography. Axelius's and Bremer's statements can be extended to most of the historical biogeographic techniques, be they cladistic or not. This is because many of them require the use of clearly delimited areas, even if these areas have not been formally defined as areas of endemism. The importance of areas of endemism has been stressed by several authors (Nelson & Platnick, 1981; Henderson, 1991), giving them a central role in the field of biogeography.

According to Michael Heads (1999), if local vicariance is not integrated into the analysis and polyphyletic areas of endemism are used to begin with, it is inevitable that the biological cladograms and the geological split sequence will be incongruent. This is because there is a tendency in the systematic work to concentrate in restricted geographic areas and see them as biogeographic regions. Even in monographic works it is traditional to place all the species of each country on separate maps. This style of presentation can easily obscure a biogeographic pattern of local vicariants, each with different intercontinental affinities.

Even if the areas of endemism constitute the operational units of most of the methods applied in historical biogeography, there exist some authors such as Peter Hovenkamp (1997) who maintain that areas of endemism should not be considered a central question of historical biogeography, and they question the existence of these areas in nature. Hovenkamp states that most biogeographic methods presuppose the existence of areas of endemism, but that there does not exist a general biogeographic theory that predicts their existence. Hovenkamp (1997) specifies that his technique, which he names "vicariance events," does not require the use of areas. He states, "The history of the earth does not resolve itself in terms of areas, but of in terms of vicariance events."

Other techniques that do not require areas of endemism as study units correspond to two variants of PAE. The first is B. Rosen's technique (1988), PAE based on localities developed for use in paleontology, and the second was developed by Morrone (1994a), presented in this book as PAE based on quadrats. In the latter case, the delimitation of areas of endemism is the goal of the method.

Panbiogeography does not always use areas of endemism, but one of its principal concepts (generalized track) shares characteristics with the area of endemism concept, as a generalized track results from the overlapping of the distribution of two or more taxa. Also studies applying the panbiogeographic techniques of track compatibility and track analysis have been developed using areas of endemism (two examples are presented in chapter 5).

Other historical biogeography techniques require the use of areas in one time or another, and even if they do not mention explicitly the area of endemism concept, it underlies all of them. Nevertheless, the definition as well as the delimitation of these areas are still controversial. In spite of the importance of this concept its critical treatment in literature is scarce, and in most biogeography texts operative definitions about areas of endemism are not given (Henderson, 1991).

Definition

In the twentieth century, scientists' approach to biogeography changed from pure description to focused analysis. The consequent development of analysis methods in the last decades of the century has made a precise understanding of the units to be studied a necessary precondition of all research (Hausdorf, 2002). However, scientists have not yet reached a consensus about how an area of endemism should be defined. Note that there exists an analogous discussion in the field of biological systematics, where for centuries scientists have tried and failed to agree on a precise definition of species (Crisci, 1981).

Many definitions have been proposed for the area of endemism concept. Nelson and Platnick (1981) define areas of endemism as fairly small areas that have a significant number of species that occur nowhere else, or else an area that is represented by more or less coincident distributions of taxa that occur nowhere else. Platnick (1991) later defined it as the congruent distributional limits of two or more species. Harold and Mooi (1994) define an area of endemism as a geographic region comprising the distributions of two or more monophyletic taxa that exhibit a phylogenetic and distributional congruence and having their respective relatives occurring in other such defined regions. Morrone (1994a) defines it as an area of nonrandom distributional congruence among different taxa. And finally, Humphries and Parenti (1999) note that an area of endemism is recognized for the coincident distribution of two or more organisms. Harold and Mooi's definition emphasizes the phylogenetic history of the taxa whose distribution patterns show congruence. These authors emphasize the historical component in the area of endemism definition, leaving aside the ecological component (Posadas & Miranda–Esquivel, 1999). However, the origin of all biogeographic patterns is never completely historical nor completely ecological, but the result of a combination of both kinds of processes (Morrone & Crisci, 1995).

Delimitation

As mentioned earlier, another question both central and controversial to the understanding of areas of endemism is their delimitation. There are diverse criteria, several of which we will discuss here (Roig-Juñent, 1999; Roig-Juñent et al., 2002): overlapping of distribution areas (Müller, 1973), area quadrating analyzed by parsimony methods (Morrone, 1994a) or by phenetic methods (Artigas, 1975), biogeographic units and overlapping of endemic taxa (Crisci et al., 2000), and partial overlapping of grouped areas (Harold & Mooi, 1994).

Overlapping distribution areas (Müller, 1973) requires the coincidence of several taxa whose taxonomic validity is unquestionable. Moreover, the distribution area of every taxon must be smaller than the area under consideration, and its limits must be clearly defined. The method consists of superposing the taxon distribution areas and establishing the overlapping area or areas, which constitute the areas of endemism.

Morrone (1994a) proposed PAE based on quadrats to identify areas of endemism (see chapter 7). Morrone's proposal consists in dividing into quadrats the region to be analyzed, and constructing a data matrix of quadrats × taxa from the taxa distributional information. In this matrix the quadrats represent the study units and the presence (coded as 1) or the absence (coded as 0) of the taxa represent the characters. Through the application of a maximum parsimony algorithm a cladogram is obtained from the matrix, where the groups of quadrats that are joined by the presence of two or more taxa are considered as areas of endemism. Finally, the taxon distributions that are sustaining each area of endemism are mapped to delimit their borders precisely. There exist similar methods in which the quadrat analysis is done with the application of similarity algorithms (for example, cluster analysis).

Finally, overlapping biogeographic units and endemic taxa is a method commonly applied in historical biogeographic studies. For example, biogeographic units (such as regions, provinces, districts) may be delimited by climatic, geological, and biotic criteria (for example, the global units defined by Takhtajan, 1986; or those in Latin America by Cabrera and Willink, 1973). Biogeographic units are also sustained with the largest possible number of endemic taxa (plants, animals, fungi, etc.) that inhabit them.

The problem with delimiting areas of endemism is similar to the aforementioned difficulties with delimiting distribution areas. It is important to emphasize that the organism distribution areas constitute working hypotheses. It will be on these hypotheses that the area of endemism, to be used as a study unit, will be delimited.

According to Roig-Juñent and colleagues (2000), there are several problems with delimiting areas of endemism. First, there is a lack of distributional information and a bias toward locality data. Points on a map represent observations, but this kind of representation makes the joint analysis of different taxa difficult. A second important difficulty concerns the congruent distributional limits of species, since the distributions of members of a diverse assemblage are usually non-sympatric. Because of this it is necessary to define limits to areas of endemism (for example, the last step in Morrone's method). This introduces a great risk of subjectivity (for example, scale distortions, sample problems, scarce data for determined taxa). On the other hand, Crisp and colleagues (1995) consider that, in general, the different kinds of areas of endemism delimitations are controversial. These authors emphasize that Morrone's method (1994a) is objective and it allows certain distribution overlapping, but question whether or not the hierarchical model of PAE is adequate to delimit areas. Linder (2001) postulated that since there are several methods for finding areas of endemism, it is desirable to have optimality criteria that allow the researcher to establish which scheme returns the best estimate of the areas of endemism for the set of species included in the analysis. These optimality criteria consider the number of species included in the areas of endemism, the number of areas delimited, and the degree of distributional congruency of the species restricted to each area of endemism.

Size

Yet another question that must be considered in relation to areas of endemism is their size. Platnick (1991) postulates that for the purposes of biogeographic study, great parts of the surface of the Earth can be termed "areas" (Australia, New Zealand, southern South America, and so on), and such treatment supposes that these regions, defined geographically more than biologically, join more than one area of endemism; an assumption that, in Platnick's view, is probably false in most cases. Nevertheless, such areas can actually constitute areas of endemism if the scale of study is large enough, and minor areas of endemism can be recognized within the larger area. Harold and Mooi define areas so broadly that there is no overlapping of distribution. In this way they avoid a problem of cladistic biogeography—widespread taxa—but their model is not applicable in certain cases. Australia is a good example of Harold and Mooi's definition, where the partial overlapping of a high number of taxa reduces the total study area (Australia) to a sole unit. On the other end, Nelson and Platnick (1981) delimit small areas. In a study that includes a high number of taxa, they define areas with few or no endemic taxa. In cladistic biogeography (see chapter 6), however, the absence of taxa in an area ("absent area") causes ambiguities in the analysis. The use of large areas of endemism permits the analysis of, for instance, relations of the biota of different continents, whereas the definition of minor areas of endemism allow the study of biota relations on the same continent.

The hierarchies of areas of endemism could be correlated with taxonomic hierarchies. In that way an area of endemism defined by the distribution of two or more species could be contained in a major area defined by the coincident distribution of two or more genera or families.

AREAS AND SAMPLING OF TAXA

Molecular phylogenies are usually generated with a sampling approach, where a taxon is considered to be a representative of a higher taxon to which it belongs. For example, the species *Smilax glauca,* native to most of eastern North America, is used as a representative of the nearly world-wide genus *Smilax* (ca. 300 spp.) in a study of the family Smilacaceae (10–12 genera). When these molecular phylogenies are used to establish biogeographic inferences (see chapter 12) there are two options for defining the distribution areas. First, one can use directly the distribution of the sampled taxon (*Smilax glauca*). Second (the most controvesial and widely used option), one can use the distribution area corresponding to the higher taxon (the genus *Smilax*). Ronquist (1996), referring specifically to his method DIVA (see chapter 8), suggested a third approach: one might try to resolve lower-level relationships within the higher taxon (the genus *Smilax*), and use this phylogenetic information to reconstruct its ancestral distribution as the distribution area.

The concept of area of endemism is controversial. It is subject to discussions from the most diverse perspectives, but at the same time it is a central topic in historical biogeography. It may be expected that in the future these discussions will clarify this concept, which represents the natural unit of historical biogeography.

2

CENTER OF ORIGIN AND DISPERSAL

IN THE SECOND HALF of the eighteenth century, Carolus Linnaeus proposed that the Garden of Eden was situated on a tropical island, the only surface emerging from the primordial sea. All organisms inhabited this island. The animals and plants that required a cold climate lived near the peak of a high mountain, and those that needed a warmer climate inhabited the plains. As the seas receded, land area increased, and animals and plants dispersed from their initial habitats to their current locations. This hypothesis was intended to explain the causes of geographical distribution of organisms that lived on Earth. In accordance with the biblical account of the Garden of Eden, Linnaeus proposed that species originated in one small area, then dispersed to other areas available for colonization.

Since Linnaeus' time, the idea of a center of origin and then dispersal has been the prevailing explanation in historical biogeography for how organisms are distributed. Darwin (1859) and Wallace (1876, 1892) considered that species originate in one center of origin, from which some individuals subsequently disperse by chance, and then change through natural selection. Darwin's and Wallace's positions on dispersal represented an important change with respect to the then-prevailing theory of immutability of species and the static view of biogeography at the time.

Darwin's ideas were strongly criticized by the extensionists—Charles Lyell, Edward Forbes, and Joseph Dalton Hooker, among others. The extensionists shared with Darwin a belief in the dynamics of dispersal, but differed in the causes that explain dispersal. They firmly held that long-distance dispersal through large and persistent barriers was a very unlikely process to explain the dynamics of distribution and related phenomena such as cosmopolitan species and disjunct distributions. They established, on the other hand, that species dispersed through land bridges and ancient continents now submerged in the oceans. The extensionists' ideas had a profound influence on the biogeography of the late nineteenth century, but at the end of that century, most geologists and biogeographers abandoned this theory. The Darwin-Wallace tradition continued through the twentieth century; among its most prominent exponents have been Matthew (1915); Mayr (1946); Darlington (1957, 1965); G. G. Simpson (1965); and Raven and Axelrod (1974).

Dispersalism is based on five basic principles (Wiley, 1981). 1) Higher taxa arise in centers of origin, where subsequent speciation occurs; 2) the center of origin of a taxon may be estimated by specific criteria; 3) the distribution of fossils is essential, because the oldest fossils are probably located near the center of origin; 4) new species evolve and disperse, displacing more primitive or plesiomorphic species toward the peripheral areas, away from the center of origin, where most derived or apomorphic species will be found; and 5) organisms disperse as widely as their abilities and physical conditions of the environment permit, so derived taxa "push" primitive taxa toward the edges of the group's range. So before we move on, we must better define the concept of center of origin.

CENTER OF ORIGIN

There have been numerous proposals of techniques to establish the center of origin of a taxon (Cain, 1944). Some of them, applied a few decades ago, seem meaningless today. Some of the more frequently used criteria for determining the center of origin are:

1. *The location of the greatest variety of forms of the taxon.* The center of origin is placed in the area where the greatest diversity of the taxon is found at present.
2. *The location of the area of greatest dominance and density of distribution.* The center of origin is placed in the area where the greatest number of individuals of the taxon are found at present.
3. *The location of synthetic or closely related forms.* The center of origin is placed in the area where the most primitive forms are found at present.
4. *The location of maximum size of organisms.* The center of origin is placed in the area currently exhibiting the maximum physical development of individuals.
5. *An increase in the number of dominant genes toward the center of origin.* The center of origin is located in the area where the individuals with the largest number of dominant genes are found today.
6. *The direction of origin indicated by the annual migration routes of birds.* The center of origin of a plant species can be found by analyzing the migration routes of the birds that disseminate that plant species.

Stanley Cain (1944) evaluated these criteria for determining centers of origin, concluding that none could be trusted independently and that some were even contradictory—for example, the location of the most primitive forms versus the location of the most advanced ones.

DISPERSALISM: A FAULTY THEORY?

Dispersalism explanations tend to reside in narrative frameworks, forming irrefutable hypotheses that do not provide a general theory to explain distributional patterns. Rather, dispersal explanations consist of unique case studies of each taxon. Panbiogeographers and cladistic biogeographers therefore consider dispersalism to be an ad hoc discipline that requires external causes to explain the patterns analyzed (Croizat et al., 1974; Platnick & Nelson, 1978; Humphries & Parenti, 1986; Grehan, 1988a). As Nelson (1978) stated, concentrating on improbable dispersals as explanations for distributions results in the "science of the rare, the mysterious and the miraculous." In addition, the acceptance of dispersal as the primary causal factor of geographical distribution creates a methodological problem: If every disjunction is explained in terms of dispersal, biogeographic patterns that result from vicariance will never be discovered. Proving the point, Craw and Weston (1984) applied the methodology of scientific research programs, developed by Lakatos (1970), to biogeographic approaches, and concluded that dispersal biogeography was not a scientific program in Lakatos's sense.

In the twentieth century numerous applications of this approach have been made, some of them regarded as less than rigorous because the center of origin was established through a dubious criterion, or through dogmatic ideas about dispersal. There are, however, at least five reasons that justify the mention of this approach: Inferring the center of origin by applying some of the criteria cited above was a common practice until the early 1980s, and is still used in some cases, though most often implicitly. Scenarios established on the basis of center of origin and direction of dispersal have been used frequently to justify the success of lines of evolution and distributional patterns of world diversity. In addition, understanding these concepts allows us to appreciate more clearly the evolution of ideas in biogeography. There are some paleontology- and geology-based hypotheses, explaining the present distribution of some group of organisms, that are supported by dispersalist explanations. Until recently, dispersal hypotheses remained difficult to test. However, phylogenetic analysis of DNA sequence data and the use of molecular clocks can provide these tests (see chapter 12). Clock calibrations that provide divergence estimates substantially smaller than those proposed by vicariant events suggest recent dispersal rather than ancient vicariance. Under a vicariance model, taxa with parallel distributions would be expected to exhibit similar amounts of genetic divergence. Conversely, dispersal may be inferred if common distribution patterns are not reflected by comparable amounts of molecular divergence (Waters et al., 2000).

CASE STUDY: INTERCHANGE OF AMERICAN MAMMALS DURING THE CENOZOIC

The paleontologist George G. Simpson described the great interchange of mammals between North and South America during the Cenozoic. His study (Simpson, 1964) provides an excellent example of the center of origin and dispersal approach as it was applied in the 1960s, when this approach was the dominant paradigm of historical biogeography. According to Simpson, the migration of animals from one region to another constitutes the most important phenomenon of historical biogeography. This phenomenon determines faunistic changes more radically than the changes produced by other causes. The geographic barriers that impede the movement of fauna and the migration or expansion routes that promote the movement of fauna constitute the fundamental geographic factors of their history.

During the Tertiary in Eurasia, North America, and South America the mammal fauna was rich, complete, and well-balanced. The areas saturated by mammal groups, however, suffered occasional invasions of new, irruptive mammal groups. When an invasion of high magnitude occurred, it produced a considerable overlapping of ecological niches. Then, the environment would become supersaturated and precipitate, or, in other words, organisms began to compete and sooner or later some of them became extinct.

An important episode of faunistic supersaturation with a further period of equilibrium occurred in the great mammal interchange between North and South America during the Upper Pliocene and the Pleistocene. According to Simpson, the composition of fauna in North and South America clearly demonstrates the existence of a land bridge between both landmasses in the Pliocene, represented by the Isthmus of Panama. Table

Table 2.1. Simpson's (1964) dispersalist hypothesis offers an explanation of the quantity and origin of North and South American terrestrial mammal families at different times, and changes in the similarity between both mammal faunas in terms of families.

	South America			North America		
	Total number of families	Natives	North American	Total number of families	Natives	South American
Recent	30	16	14	23	20	3
Pleistocene	36	23	13	34	26	8
Middle Pliocene	25	24	1	27	26	1
Middle Miocene	23	23	0	27	27	0

2.1 shows the increase in biodiversity that occurred during this exchange, and how the landmasses re-established a new equilibrium. This process was accompanied by a noteworthy change in the fauna composition, especially in South America.

Some of the original postulates of Simpson's work were later refuted, such as the extinction of South American mammals because of competition, or the idea that all the biogeographic particularities of the mammals' history are best explained with the theory that the continents had the same identity and position in the past that they have today. However, there is much evidence that supports the idea of a North American origin for many South American mammals (for example, *Felis, Tapirus, Tayassu*), and that dispersal of these ancestral taxa occurred from the Northern Hemisphere through the Isthmus of Panama during the Pliocene. Reciprocally, numerous South American taxa (for example, *Didelphis, Dasypus, Eremotherium*) dispersed from South America to North America (Fig. 2.1). This process is widely known as the Great American Biotic Interchange. Supporting evidence is provided by the fact that before the interchange there is a complete absence of fossils of the immigrant groups' ancestors on the continent that they colonized. For example, there is not a registered presence of Edentata prior to the Pliocene in North America, nor are

FIGURE 2.1. Great American Biotic Interchange. These are some examples of mammals that crossed the Panama Isthmus, northward and southward respectively, during the Upper Pliocene.

there records of Felidae before the Pliocene in South America (Stheli & Webb, 1985; Marshall & Cifelli, 1990; Pascual et al., 1996). Furthermore, there is geographic evidence that supports the existence of a land bridge between North and South America during the Pliocene, at the time of the faunistic interchange (Marshall & Sempere, 1993).

3

PHYLOGENETIC BIOGEOGRAPHY

PHYLOGENETIC BIOGEOGRAPHY (Hennig, 1950, 1966; Brundin, 1966, 1981) was the first approach to consider an explicit phylogenetic hypothesis (a cladogram) of a given group of organisms as the basis for inferring its biogeographic history. This approach can be defined as the study of the history of monophyletic groups in time and space. Hennig (1966) stated that there is a close relationship between a species and the space it occupies. According to Hennig, each group of organisms has unique dispersal patterns and an independent history. The first applications of Hennig's ideas to real taxonomic groups may be found in Brundin (1966, 1972, 1981) and Ross (1974).

RULES AND METHODOLOGY

Phylogenetic biogeography applies two basic rules, the progression rule and the deviation rule. The progression rule states that primitive members of a taxon are found closer to its center of origin than more apomorphic ones, which are found on the periphery. Hennig (1966) conceived that speciation was allopatric, involving peripheral isolates, and causally connected to dispersal. Within a continuous range of different species of a monophyletic group, the transformation series of characters run parallel with their progression in space, such that the youngest members would be on the geographic periphery of the group. The progression rule is based on the peripheral isolation allopatric mode of speciation, so it cannot be applied when other modes of speciation are considered, because it is rejectable a priori.

The deviation rule states that in any speciation event, an unequal cleavage of the original population is produced, in which the species that originates near the margin is apomorphic in relation to its conservative sister species, which are more plesiomorphic. In each event of speciation, the peripheral sister species develops more evolutionary novelties, whereas the other remains closer to the ancestor.

The methodology of phylogenetic biogeography can be summarized as follows:

1. A cladogram is constructed of the group under study;
2. the areas inhabited by the group are optimized onto the cladogram;
3. the center of origin of the group is inferred through the application of progression and/or deviation rules, and the dispersal direction is determined;
4. a hypothesis of the biogeographic history of the group is formulated; and
5. this hypothesis is confirmed by matching it against the geology of the area.

Although phylogenetic and dispersal biogeography may be lumped into the same approach, because both emphasize centers of origin and dispersal, some authors (Wiley, 1981; Humphries & Parenti, 1986) regard phylogenetic biogeography as an advance over dispersalism because of the explicit use of phylogenetic hypotheses instead of descriptive enumerations and scenarios. Phylogenetic biogeography provides a more rigorous methodology because it optimizes areas onto the phylogeny of a group, and infers the fewest possible areas of dispersal for each group. Critiques of this method, however, note that it requires ad hoc assumptions about the center of origin, and that species migrate from the center of origin to other areas. Furthermore, according to Forey and colleagues (1992) and Humphries and Parenti (1999), interpreting individual cladograms as having individual histories leads to certain conceptual difficulties. One of these difficulties is the repetition of distribution patterns. For instance, if there are many distantly related taxonomic groups repeating a pattern of distribution among continents, for example, South America, Australia, and New Zealand, it is improbable that each taxon has a unique dispersal history from one continent to the other. The most logical and simplest conclusion would be that at one time, the continents were in contact and that the present-day pattern is caused by the breakup of a formerly continuous austral biota.

It is interesting to note that phylogenetic biogeography has, in part, continuity in a modern approach such as ancestral areas (see chapter 4). The idea of a center of origin located in the most plesiomorphic branches of a cladogram, one of the basic postulates of phylogenetic biogeography, is evoked by one of the criteria to determine the ancestral area (in this case ancestral area is the same concept as center of origin, though it goes by a different name).

CASE STUDY: THE CARABID BEETLES OF AUSTRALASIA

Some authors, such as Darlington (1970) in one of his studies on carabid beetles (Insecta: Coleoptera) of New Guinea, postulated that a number of these insects had dispersed (one way or another) between Asia and Australia. Brundin performed a theoretical analysis of this idea (Brundin, 1972) on the basis of phylogenetic biogeography. On the map of Australasia in Figure 3.1a each black dot marks the occurrence of a hypothetical endemic species of carabid. The species are supposed to be members of a group A + B + C + D + E that is strictly monophyletic. According to the phylogenetic biogeographic approach, this distribution picture

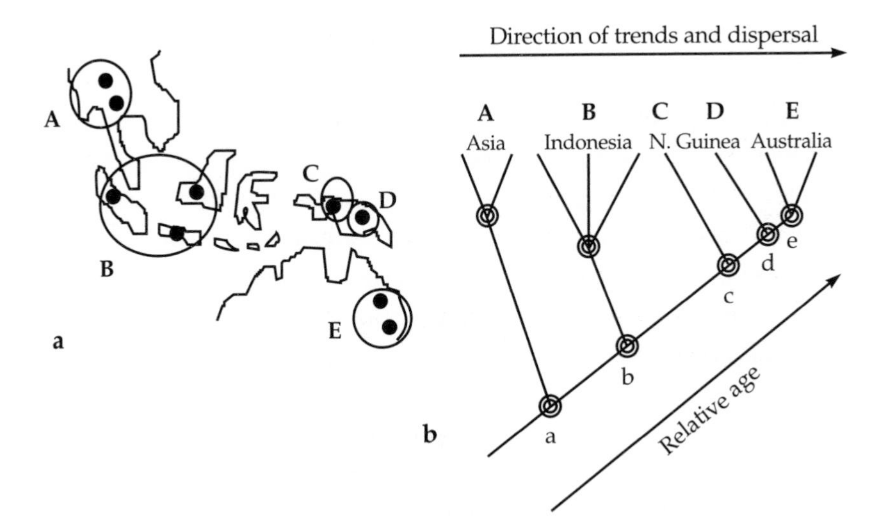

FIGURE 3.1. Phylogenetic biogeography. *(a)* Map of Australasia showing the distribution of hypothetical species A, B, C, D, and E; *(b)* phylogenetic relation of species A–E, showing step by step dispersal from southeast Asia via Indonesia and New Guinea to Australia.

is the result of step-by-step dispersal from southeastern Asia via Indonesia and New Guinea to Australia. Consequently, the phylogenetic diagram of this group shown in Figure 3.1b is in accordance with the supposed dispersal event. If so, the Australian group E would be the most apomorphic group, because it is peripheral and a member of the youngest sister group, pair D + E, in a hierarchic sequence of successively subordinate groups. The nodes a, b, c, d, and e signify a sequence of decreasing age and increasing apomorphy. According to Brundin, this example has to be considered as an expression of the ever-present parallelism between morphological and chorological progression. Dispersal, seen in the time perspective, is a multiple process including progression in space, evolutionary change, and speciation. This is in turn a consequence of the norm that speciation by cleavage of a stem species is a product of time, progression in space, and isolation of peripheral populations.

Brundin applied the phylogenetic biogeographic approach to real data. His studies (Brundin, 1966, 1972, 1981) on Podonominae chironomid midges inhabiting the southern temperate areas of South America, southern Africa, Tasmania, southeast Australia, and New Zealand are widely known. Brundin first produced a cladogram of the evolutionary relationships of the chironomid species, and in place of the name of each species in the cladogram he inserted the name of the continent in which it is found. As a result, he hypothesized that the African genera diverged first, and that the divergence of the New Zealand species preceded the divergence between the South American and the Australian species.

RESEARCH USING PHYLOGENETIC BIOGEOGRAPHY

Recent examples of this approach consist mainly of papers that explicitly or implicitly apply this method or that employ its assumptions in part. Among them, we can cite that of Knox and Palmer (1998) on Lobeliaceae from eastern Africa.

4

ANCESTRAL AREAS

KÅRE BREMER (1992, 1995) formalized a cladistic procedure based on the dispersalist view. This author considered that understanding ancestral areas for an individual group is a valid part of the study of the natural history of that group, and that the previous approach, searching for a center of origin, not the search per se, was spurious. Bremer's procedure allows the researcher to identify the ancestral area of a group from the topological information of its cladogram.

Those who use this method make two assumptions. First, areas that are positionally more plesiomorphic (present on deep branches) in a cladogram of a particular group are more likely parts of the ancestral area for that group than those areas that are positionally more apomorphic. Second, areas represented on numerous branches of the cladogram are more likely parts of the ancestral areas than are areas represented on few branches. These assumptions are based on two criteria for determining the center of origin and dispersal (see chapter 2). The first assumption is based on the location of synthetic or closely related forms, whereas the second assumption is based on the location of the greatest variety of forms of the taxon.

CAMIN-SOKAL OPTIMIZATION

As proposed by Bremer (1992), the first step for analyzing ancestral areas is to choose a cladogram or consensus tree of the group using standard phylogenetic inference techniques (Fig. 4.1a). The next step is to determine the individual distribution areas to be used in the analysis (Fig. 4.1b). Each area can be considered as a binary character with two states (present or absent). This method applies the irreversible Camin-Sokal parsimony algorithm (see Appendix A) to quantify two parameters: the depth of the areas in the cladogram (the distance with respect to the basal nodes), and the number of times that each area is present in the cladogram.

The method starts with the assumption that the ancestral area is identical to the present areas of distribution of taxa. Under this all-loss/no-gain model, the ancestral state for each character is present (= 1) with the assumption of 1 → 0 irreversibility (Figs. 4.1c–f). All area absences are plotted as losses onto the cladogram (indicated with crosses). Then, the alternative is considered, in which the ancestral area is empty, and none of the individual areas are part of the ancestral area. In this model there are no losses and all area presences are the result of gains. The ancestral state is specified as absent (= 0) with the assumption of 0 → 1 irreversibility. The necessary gains are indicated by bars (Figs. 4.1c–f).

Finally, the number of gains (G) and the number of losses (L) for each individual area are compared (Fig. 4.1g). The gain/loss (G/L) quotient may be used to compare the relative probabilities that individual areas were indeed part of the ancestral area. A high value G/L for an individual area indicates a higher probability that the area was part of the ancestral area. By rescaling the G/L quotients to a maximum value of 1, the values are more easily compared. Rescaled quotients (AA) are obtained by dividing by the largest G/L quotient. Figure 4.1g shows that areas I and II (both with G/L = 1) are most likely parts of the ancestral area for the hypothetical example of taxa A–D.

The quotients allow a comparison between gains and losses. Initially,

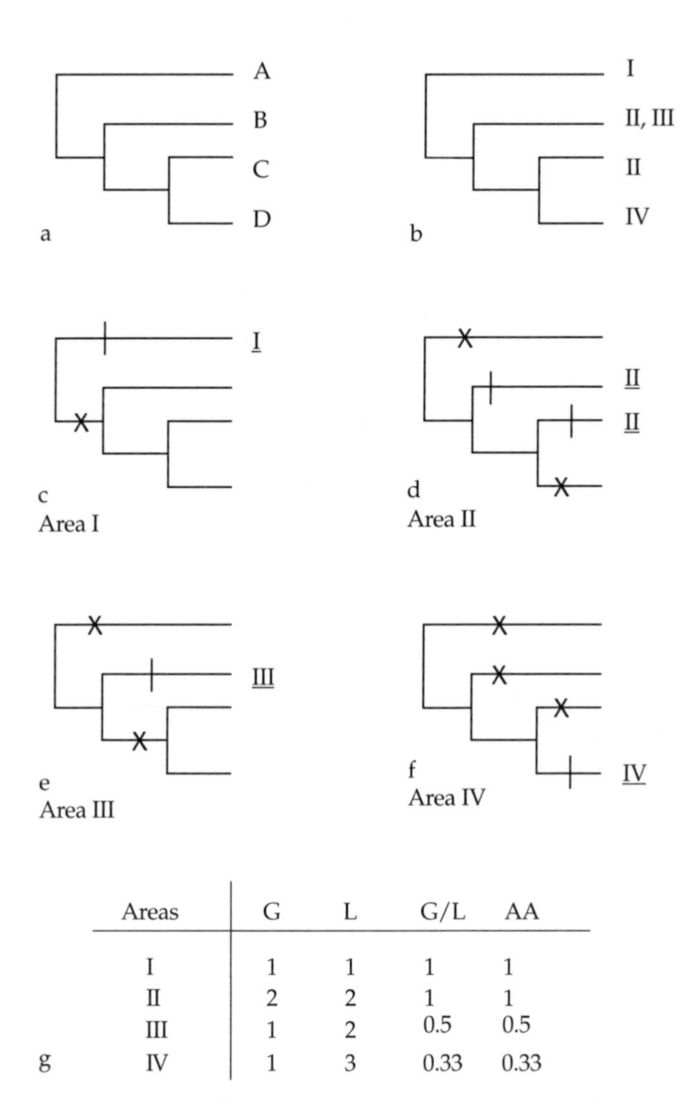

Areas	G	L	G/L	AA
I	1	1	1	1
II	2	2	1	1
III	1	2	0.5	0.5
IV	1	3	0.33	0.33

FIGURE 4.1. Application of the ancestral areas method using Camin-Sokal parsimony. *(a)* Cladogram of species A–D; *(b)* cladogram of areas I–IV; *(c–f)* gains (bars) and losses (crosses) using Camin-Sokal parsimony for the four areas I, II, III, and IV; *(g)* ancestral areas estimation for species A–D. G = number of gains required under Camin-Sokal parsimony. L = number of losses required under Camin-Sokal parsimony. AA = G/L rescaled quotients to a maximum value of 1 by dividing with the largest G/L value.

gains and losses are assumed equally possible. Then, if there are more losses than gains for any individual area, the all-loss interpretation for this area is rejected as less parsimonious, and the area is excluded from the ancestral area. However, if there are more gains than losses for any individual area, the all-gain interpretation for this area is rejected, and the area is assumed to be part of the ancestral area.

FITCH OPTIMIZATION

Fredrik Ronquist (1994, 1995) proposed a method that uses Fitch optimization or reversible parsimony, essentially as a critique of Bremer's method (1992). According to Ronquist, Bremer's method is flawed for several reasons:

It uses forward and reverse Camin-Sokal parsimony instead of a reversible parsimony (for example, Wagner parsimony, Fitch parsimony);

it employs Camin-Sokal (irreversible) parsimony in reconstructing ancestral areas, producing meaningful results only if the process itself is irreversible (in other words, it assumes that dispersal always occurred going away from the ancestral areas);

polymorphisms are not considered in Bremer's method. When some of the recent species occur in more than one area, it is possible that the ancestors also occurred in more than one area;

multistate characters are not considered in Bremer's method. When the recent distributions involve only mutually exclusive areas of endemism (comparable to a multistate morphological character without polymorphisms), the ancestral areas can be reconstructed using Fitch optimization for unordered states. If the geographic position of the areas suggests a particular dispersal sequence, the area states may be ordered in that sequence.

The differences between Bremer's and Ronquist's methods can be seen in the following example, where both results show different ances-

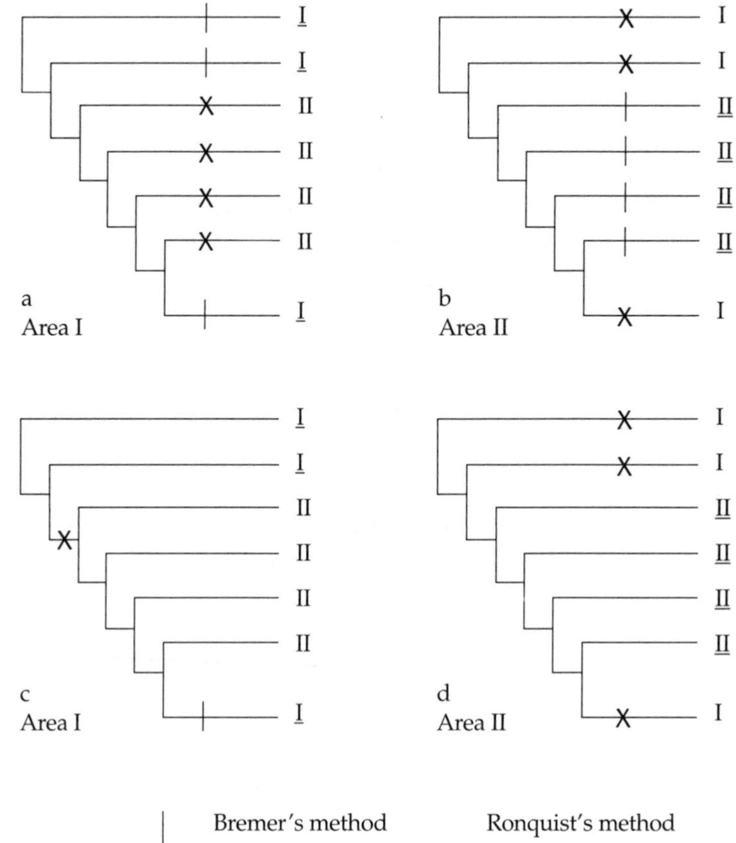

e

Areas	Bremer's method G	L	G/L	AA	Ronquist's method S	RP
I	3	4	0.75	0.56	2	1
II	4	3	1.33	1	3	0.67

FIGURE 4.2. Ancestral areas. Comparison of Bremer's (using Camin-Sokal parsimony) and Ronquist's (using Fitch parsimony) methods. *(a–b)* Areas I and II optimization using Camin-Sokal parsimony; *(c–d)* Areas I and II optimization using Fitch parsimony; gains are represented by bars and losses are represented by crosses; *(e)* table summarizing both method results. G = number of necessary gains under forward Camin-Sokal parsimony; L = number of losses under reverse Camin-Sokal parsimony; G/L = gain/loss quotient; AA = G/L rescaled quotients to a maximum value of 1 by dividing with the largest G/L value; S = number of necessary steps if the area was the ancestral area; RP = S values rescaled to a maximum value of 1 by inverting them and multiplying by the smallest S value.

tral areas for the same cladogram. Bremer's method (Fig. 4.2a–b) indicates that area II (AA = 1; Fig. 4.2e) is more likely to be the ancestral area than area I (AA = 0.56). Under reversible parsimony (Fig. 4.2c–d), however, the ancestral area assignment is I because if the ancestral area was II, three steps *(S)* would be required to explain the character, whereas only two steps are required if it was area I. To obtain a measure of the relative probabilities of the ancestral area being I or II under reversible parsimony, *S* values are inverted and multiplied by the smallest *S* value. For example, the relative probability for I is $1/2 \times 2 = 1$, and for II, $1/3 \times 2 = 0.67$ (Fig. 4.2e).

WEIGHTED ANCESTRAL AREA ANALYSIS

Bernhard Hausdorf (1998) proposed another method for estimating ancestral areas. This method is based on reversible parsimony in combination with a weighting scheme that weights more highly steps in positionally plesiomorphic branches than steps in positionally apomorphic branches, thus returning to the "location of synthetic or closely related forms" criterion for determining centers of origin or ancestral areas.

This method develops a new algorithm for ancestral area analysis that avoids the aforementioned drawbacks of Bremer's (1992) and Ronquist's (1995) methods. For example, in the latter the number of steps *(S)* depends on increasing tree size. Hausdorf proposes a parsimony technique permitting reversible changes between the states (in this case the areas), as recommended by Ronquist. However, in contrast to Ronquist's method, the most probable reconstruction for the ancestor is not always the state that, given the tree, requires the minimum number of steps to explain the distribution of states among terminal taxa. Loss and gain steps have opposite influences on the calculation of the probability that an area was part of the ancestral area. A step in which area *X* is replaced by a different area diminishes the probability that area *X* was part of the ancestral area. However, if area *X* is recolonized by a subgroup of that branch, this step increases the probability that this area was part of the ancestral area. Therefore, gain steps and loss steps concerning area *X* are counted separately, much as they are in Bremer's method. An additional modification is necessary to prove that the areas that are positionally plesiomorphic in the area cladogram are more likely parts of the ancestral areas. Hausdorf's method weights steps in the positionally plesiomorphic branches more heavily than those in the positionally apomorphic branches. A simple concave function $1/x$ is used, in which x is the number of internodes as counted from the common ancestor.

In summary, the method for calculating an index for the probability that an individual area was part of the ancestral areas follows these steps (Hausdorf, 1998):

1. Optimize area *X* (character *X*) onto the cladogram, under the assumption that area *X* was not part of the ancestral area;
2. weight the steps in the branches above the x internode with $1/x$ and count the number of weighted gain steps (GSW);
3. optimize area *X* onto the cladogram under the assumption that area *X* was part of the ancestral area;
4. weight the steps as described and count the number of weighted loss steps (LSW);
5. repeat steps 1–4 for every area; and
6. calculate the probability index, PI = GSW/LSW, for every area.

In the cladogram in Figure 4.2, if area I was not part of the ancestral areas, there are three gain steps located in the first, second, and last branch of the cladogram, and one, two, and six nodes with respect to the ancestral node x. Therefore,

$$\text{GSW} = 1/1 + 1/2 + 1/6 = 1 + 0.5 + 0.17 = 1.67$$

If area I was part of the ancestral areas, there is only one loss step with two internodes from this step to the ancestral node: LSW = $1/2 = 0.5$. Repeating the calculations for area II, GSW = 0.5 and LSW = 1.67. The probability index is PI = 3.34 for area I and PI = 0.30 for area II, indicating that area I, with the highest value, is more likely to be part of the ancestral area.

According to Hausdorf, this method is also appropriate for the solution of the redundant distribution problem in area cladograms. Under the assumption of allopatric speciation, redundant distributions (sympatry of sister groups) show that dispersal has occurred. Thus, the ancestral area of at least one sister group was smaller than the combined distribution of its descendants. With the weighted ancestral area analysis, the ancestral areas can be confined and at least some dispersal events can be distinguished from possible vicariance events. One of the problems of this approach is that the PI is infinite for all areas of terminal branches. This may result in an overestimation of the ancestral area of widespread terminal taxa, and an underestimation of the ancestral area of their sister groups.

Malte Ebach (1999) has criticized ancestral area methods because they use, and do not reduce, paralogy to find the center of origin.

CASE STUDY: ANCESTRAL AREAS OF *MOSCHARIA* AND *POLYACHYRUS* (ASTERACEAE)

The genera *Moscharia* (two species) and *Polyachyrus* (seven species) form a monophyletic group within Mutisieae, a phylogenetically basal tribe of Asteraceae. Both genera are distributed in the Andean and coastal dry areas of Chile and Perú, from 8° to 35°S in South America. These areas of Chile and Perú are of biogeographic interest because of their high endemism and the competing hypotheses that have attempted to explain distributions of taxa in this restricted area. Therefore, it is worthwhile to estimate the probable location of an ancestral geographic area for *Moscharia* and *Polyachyrus* applying Bremer's method (Katinas & Crisci, 2000).

Area delimitation: Areas of endemism were circumscribed to include several endemic taxa, following the studies of Cabrera and Willink (1973) and Morrone and colleagues (1997). The areas considered in this study were the Coastal Desert, Cardonal, North Central Chile, and South Central Chile.

Construction of taxon cladogram: A most parsimonious cladogram was obtained on the basis of morphological characters, using standard cladistic techniques.

Construction of taxon-area cladogram: The taxon-area cladogram was constructed by replacing the names of the terminal taxa with the names of the areas in which they occur. Each area is treated as a single character, which was optimized onto the taxon-area cladogram using either forward or reverse Camin-Sokal parsimony (Fig. 4.3a–d).

Identification of ancestral area: The number of gains or presences (G) and losses or absences (L) for each individual area was computed (Fig. 4.3e). The G/L ratio for each geographic area was estimated to find the individual areas with the highest G/L ratios. These could be deemed a part of the hypothetical ancestral area (AA). The ratios G/L obtained were: Coastal Desert = 1, Cardonal = 0.5, North Central Chile = 1.3, and South Central Chile = 1. These results indicate that the North Central Chile area could be a part of the ancestral geographic range of *Moscharia* and *Polyachyrus,* given its relatively high G/L value.

Different geological events in southern South America may have played an important role in the evolution of the biota of this area, including *Moscharia* and *Polyachyrus.* Among these events, the uplift of the Andes in the late Oligocene, and the Quaternary glaciation through a series of wet-dry cycles may have led to the present distribution of *Moscharia* and *Polyachyrus* and other taxa. It can be hypothesized that the ancestor of *Moscharia* and *Polyachyrus* may have inhabited part of North Central Chile (30°–35°S). During humid climate periods, the biota inhabiting this region increased its range both to the south (South Central Chile area) and to the north (Coastal Desert area), with the high Andean slopes (Cardonal area) being the last area to be occupied.

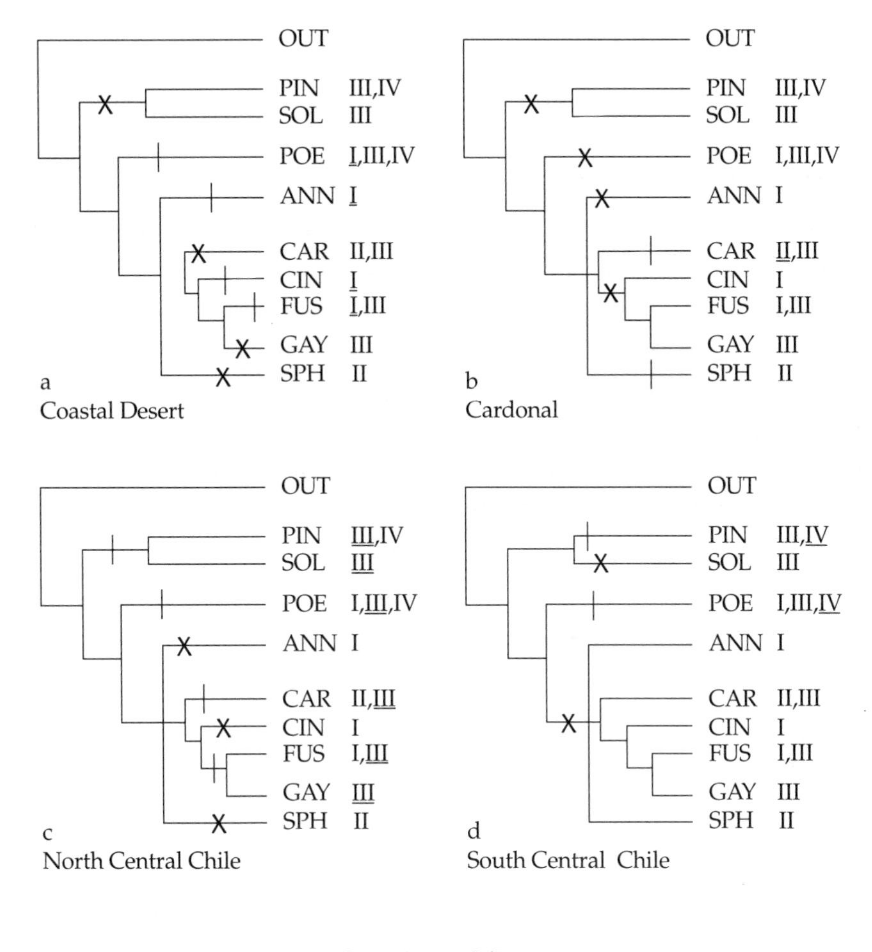

e

Areas	G	L	G/L
I	4	4	1
II	2	4	0.5
III	4	3	1.3
IV	2	2	1

FIGURE 4.3. Ancestral areas. *(a–d)* Area cladograms of *Moscharia* and *Polyachyrus.* Gains (bars) and losses (crosses) for the four areas according to Camin-Sokal parsimony are indicated. *(a)* Coastal Desert; *(b)* Cardonal; *(c)* North Central Chile; and *(d)* South Central Chile. *(e)* Table summarizing gains (G), losses (L), and G/L quotients for each area. OUT = outgroup, ANN = *Polyachyrus annus,* CAR = *P. carduoides,* CIN = *P. cinereus,* FUS = *P. fuscus,* GAY = *P. gayi,* PIN = *Moscharia pinnatifida,* POE = *Polyachyrus poeppigii,* SOL = *Moscharia solbrigii,* SPH = *Polyachyrus sphaerocephalus.*

RESEARCH USING ANCESTRAL AREAS

There are some empirical applications that use the ancestral areas approach. Among them, we can mention Ezcurra and colleagues (1997) on the plant genus *Chuquiraga* from South America; Repetur and colleagues (1997) on the plant genus *Bromheadia* from southeast Asia; Swenson and Bremer (1998) on the plant genus *Abrotanella* from the Southern Hemisphere; Fritsch (1999) on the widely distributed plant genus *Styrax;* Voelker (1999a) on the cosmopolitan Passerine genus *Anthus;* and Hausdorf (2000) on the widely distributed Gastropods Limacoidea.

5

PANBIOGEOGRAPHY

PANBIOGEOGRAPHY, originally proposed by the French-Italian botanist Léon Croizat (1952, 1958, 1964, 1981) and recently revised and updated by Robin Craw and colleagues (1999), is an approach to biogeography that focuses on the role of locality and place in the history of life. Central to panbiogeography is the acknowledgment that an understanding of locality is a fundamental precondition to any adequate analysis of the patterns and processes of evolutionary change. Furthermore, for panbiogeographers location is potentially more important than any other factor, such as ecology or isolation, in determining the characteristics of a biota. To exemplify this viewpoint, Michael Heads (pers. comm.) states that the flora of New Caledonia is often attributed to long isolation or ecology (variety of habitats), but probably neither factor has much to do with generating the flora in the first place, even though they may permit survival. As expressed by some biogeographers, in panbiogeography, as in real estate, there is only location, location, location.

Croizat (1964) postulated that "earth and life evolve together," and that "life is the last geological layer," by which he meant that barriers evolve together with biotas. From this metaphor grew the concept of panbiogeography. Croizat plotted distributions of organisms on maps and connected the disjunct distribution areas or collection localities with lines called tracks. Individual tracks for unrelated groups of organisms were then superimposed, and if they coincided, the resulting summary lines were considered generalized tracks. Generalized tracks indicate the preexistence of ancestral biotas, which subsequently become fragmented by tectonic and/or climatic change. When two or more generalized tracks intersect in an area, that area is called a node.

Generalized tracks commonly stretch across the major oceanic barriers (for example, between New Zealand and South America). Croizat thus developed the concept of considering the ocean basins, not continents, as the major focus of biogeographic attention (Keast, 1991).

Attempts to develop a quantitative, repeatable, and statistical basis for track analysis were pioneered by McAllister and colleagues (1986), Page (1987), Connor (1988), Craw (1989), and Henderson (1990, 1991). These attempts at track analysis used principles and practices derived for the most part from graph theory to provide objective and quantitative methods for drawing and comparing tracks. One of these methods is the well-accepted track compatibility developed by Craw (1988a, 1989). Another procedure, proposed by Page (1987), is based on graph theory but has not been yet applied to real data.

TRACK ANALYSIS

Panbiogeography involves three principal concepts: individual track, generalized track, and node (Craw, 1979, 1983, 1984a,b, 1985, 1988a,b; Page, 1987; Craw & Page, 1988; Grehan, 1991, 2001a; Crisci & Morrone, 1992b; Morrone & Crisci, 1995; Craw et al., 1999).

An individual track represents the spatial coordinates of a species or group of related species, and operationally is a line graph drawn on a map of their localities or distribution areas, connected according to their geographical proximity. In graph theory (Wilson, 1983), a track is equivalent to a minimal spanning tree, which connects all localities to obtain the smallest possible link length (Fig. 5.1). After the track is constructed, its

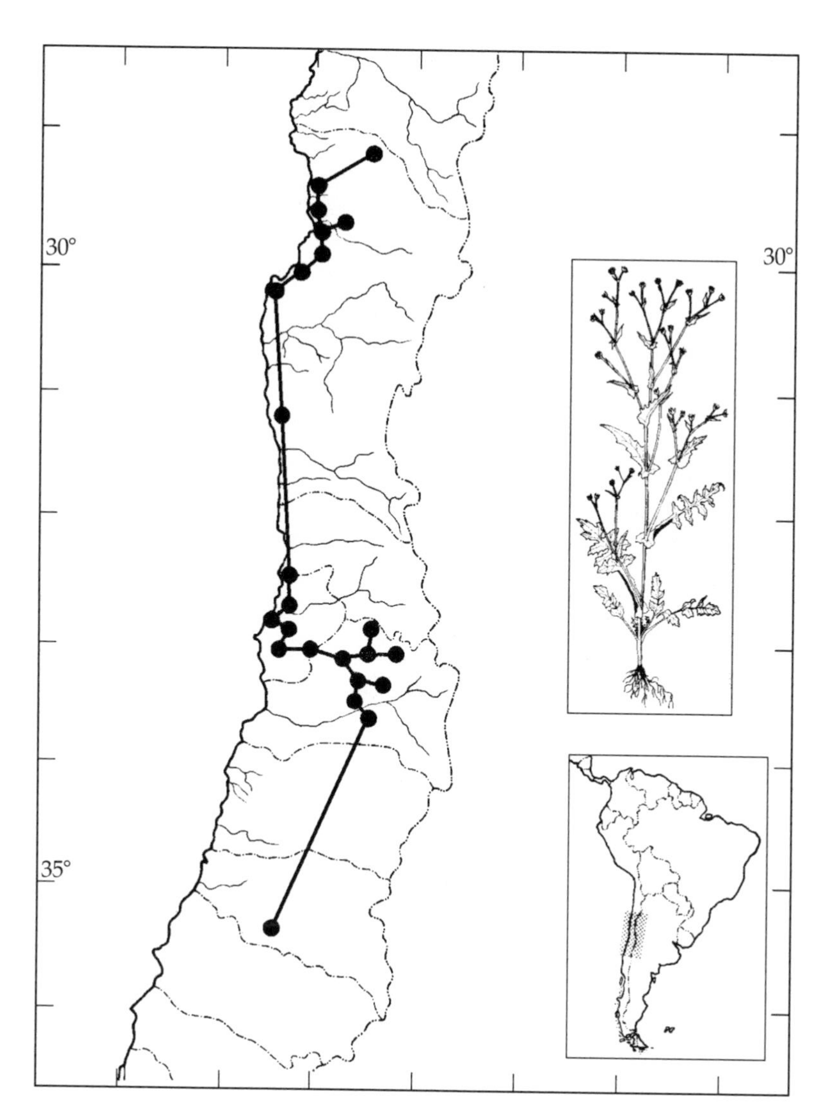

FIGURE 5.1. Individual track, habit, and distribution of *Moscharia pinnatifida* (Asteraceae). The localities of *M. pinnatifida*, placed on the Central Chile area, are connected according to their geographic proximity, constituting an individual track equivalent to a minimum spanning tree (modified from Crisci et al., 2000).

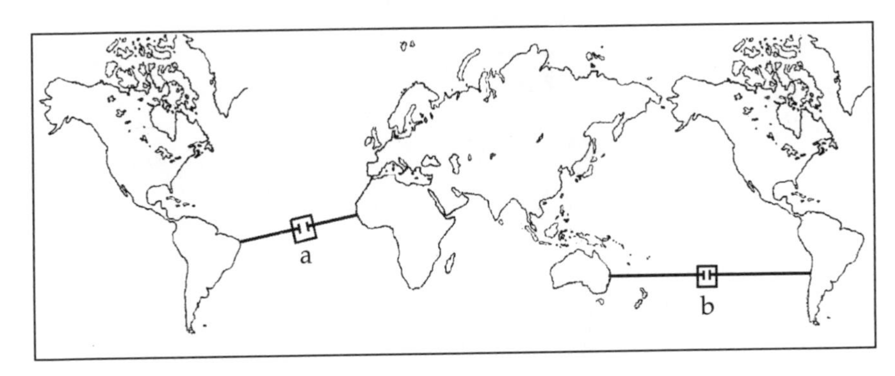

FIGURE 5.2. Baseline. World map showing an Atlantic baseline (a) and a Pacific baseline (b).

orientation (rooting) can be determined using one or more of the following criteria:

Baseline: A baseline is a spatial correlation between a track and a specific geographic, geological, or geomorphic landscape feature, such as the crossing of an ocean or sea basin, or a major tectonic structure, interpreted as a diagnostic character (meaning spatial homology) (Fig. 5.2). Track polarity is determined by orienting the track from the baseline to the other localities connected by it.

Main massing: A main massing is the greatest concentration of biological diversity within the geographic range of a taxon. This diversity may be measured by taxonomic diversity (for example, number of genera, species, or subspecies), genetic, phenotypic, or behavioral characters. Track polarity is determined by orienting track links from localities more proximate to the main massing to those that are more distant.

Phylogeny: If cladistic information is available, it can be used to direct the track from the most primitive to the most derived taxa. Main massing and phylogeny recall concepts of centers of origin and dispersal, just as in the ancestral area method.

Individual tracks for unrelated taxa or groups that coincide constitute a generalized or standard track. It provides a spatial criterion for biogeographic homology (Grehan, 1988b), indicating the preexistence of ancestral biotas that became fragmented by physical or geological events. A generalized track is a set of two or more individual tracks that are compatible or congruent according to a specified criterion (for example, shared baselines or compatible track geometries).

A node is the area where two or more generalized tracks overlap. It means that different ancestral biotic and geologic fragments interrelate in space-time, as a consequence of terrane collision, docking or suturing, thus constituting a composite area. The nodes are dynamic biogeographic boundaries where remnant fragments of different ancestral biotas came into contact.

The panbiogeographic approach may be exemplified by analyzing three austral taxa: Ratites (Aves), *Nothofagus* (Nothofagaceae), and *Leiopelma* (Amphibia) (Fig. 5.3). Their individual tracks show that these taxa do not share spatial homology. Only the Ratites are clearly Gondwanic, having their distribution oriented by the basins of the Atlantic and Indian Oceans. In spite of partial sympatry in Australia and southern South America, only *Leiopelma* and *Nothofagus* are geographically homologous, belonging to the same ancestral biota, which is different from that of the Ratites. This result contrasts with biogeographic studies in which Ratites and *Nothofagus* have been assumed a priori to belong to the same ancestral biota.

The panbiogeographic approach has been subject to several criticisms (Mayden, 1991). Patterson (1981) and Seberg (1986) have claimed that panbiogeographers use systematic treatments in an uncritical way. Main massing has been considered similar to centers of origin, and therefore has been subject to the same criticism (Platnick & Nelson, 1988; Humphries & Seberg, 1989). This similarity between the main massing and centers of origin concepts is rejected by panbiogeographers. Michael Heads (pers. comm.) argued that just because tracks are oriented with different polarity does not mean there has been any migration by physical movement. Platnick and Nelson (1988) rejected the use of geographical proximity for drawing tracks because they considered that

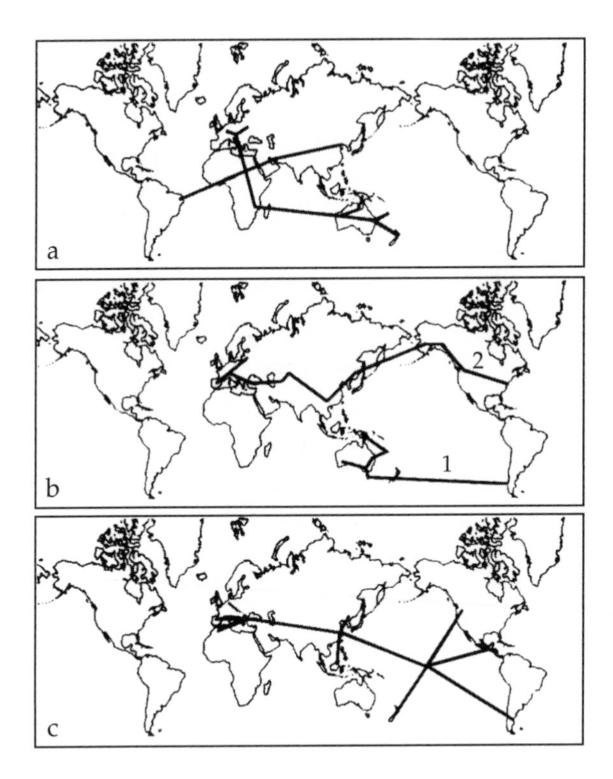

FIGURE 5.3. Individual tracks showing different baselines. *(a)* Ratites (Aves); *(b1) Nothofagus* (Nothofagaceae, Southern Hemisphere); *(b2) Fagus* (Fagaceae, Northern Hemisphere); and *(c) Leiopelma* (Amphibia) and related taxa.

cladistic information is a prerequisite to any historical biogeographic analysis.

It is important to note that a methodology called "continuous track analysis" or CTA (Alroy, 1995) also makes use of the term "track," although it has no relation to panbiogeography. CTA attempts to depict reticulate evolutionary patterns in phylogenetics and biogeography.

TRACK COMPATIBILITY

The track compatibility method was developed by Craw (1988a). It is based on the concept of distributional compatibility (Connor, 1988; Craw,

1989). In this method, individual tracks are treated as biogeographic hypotheses of locality or distribution area relationship. Two or more individual tracks are regarded as being compatible with each other only if they are the same or if one track is a subset of the other, i.e., tracks are either included within or replicated by one another.

The track compatibility method analyzes the data in a way analogous to character compatibility (Meacham, 1984). However, the track compatibility method uses the concept of compatibility in a restricted way compared to its original usage in taxonomy. For example, in the track compatibility method, nonoverlapping tracks are incompatible, though they would be compatible under the original taxonomic concept.

The method consists of constructing a matrix (areas versus tracks) in which each entry is 1 or 0 depending on whether the track is present or absent in each area respectively, and using a compatibility analysis program to find the largest cliques of compatible tracks. The tracks are treated as binary characters ordered with absence as the "ancestral" state for each (= all zero outgroup). The method involves finding a simple form of spanning tree linking localities or distribution areas. The tree is constructed from the largest clique of compatible distributions in a distributional compatibility matrix and is based on the original concept of compatibility (nonoverlapping tracks are also considered compatible). Therefore, using a restricted concept of compatibility (only individual tracks that are either included within or replicated by one another are compatible) the tree (= clique) could contain more than one generalized track. In this case the multiple generalized tracks (groups) found in one clique will be formed only by areas that are exclusive of each generalized track. If more than one large clique or several cliques of considerable size are found, then it may be hypothesized that several generalized tracks link the localities or distribution areas in more than one way (in this case, areas or localities can be considered members of more than one generalized track at the same time). Alternatively, the intersection (those tracks common to all the largest cliques) of the largest cliques can be identified as a generalized track (Craw, 1990).

In Figure 5.4, there are four individual tracks (A, B, C, and D). The

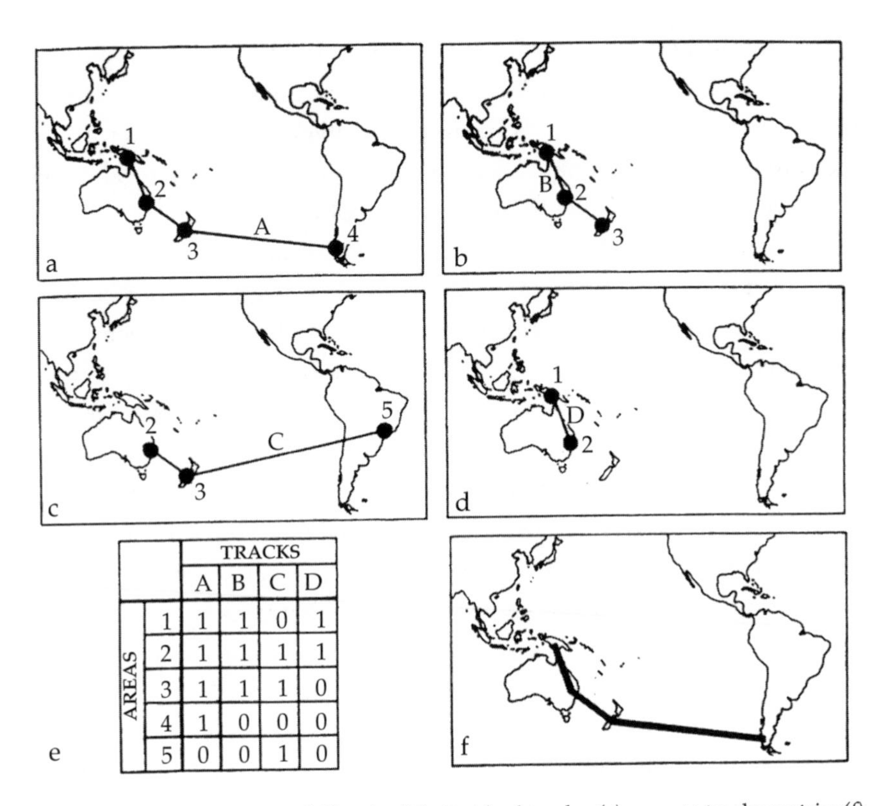

	TRACKS			
AREAS	A	B	C	D
1	1	1	0	1
2	1	1	1	1
3	1	1	1	0
4	1	0	0	0
5	0	0	1	0

FIGURE 5.4. Track compatibility. *(a–d)* Individual tracks; *(e)* areas × tracks matrix; *(f)* generalized track. Individual tracks are indicated by letters: A, B, C, and D, and areas by numbers: 1, 2, 3, 4, and 5.

matrix of areas × tracks, analyzed with a compatibility algorithm, produces a generalized track based on tracks A, B, and D, with C incompatible with them. For a track compatibility analysis, the CLIQUE computer program in the PHYLIP package (Felsenstein, 1993) and SECANT version 2.2 (Salisbury, 1999), based on Fiala's (1984) program CLINCH, can be used. For more details and other applications of this method see Craw (1988a, 1989); Morrone and Crisci (1990, 1995); Craw and colleagues (1999); and Crisci and colleagues (2000).

CASE STUDY 1: TRACK ANALYSIS IN THE ANDEAN SUBREGION

Some authors (for example, Crisci et al., 1991a) postulated that the South American biota has a composite origin, with an austral component related to the biota of the Austral areas and a tropical component related to the African biota. Furthermore, it was postulated that temperate South America might be a composite area in itself. A track analysis of the Andean subregion was developed in order to elucidate further the nature of the temperate areas of South America (Katinas et al., 1999).

Area delimitation: The Andean subregion (Fig. 5.5a) (Morrone, 1994b, 1996a) corresponds to southwestern South America below 30°S, also extending through the Andean highlands northward of this latitude to Venezuela. It comprises five provinces, delimited by several endemic taxa: Subantarctic, Central Chilean, Patagonian, Puna, and Páramo. The biogeographic regions of the world were also considered: Holarctic (Palearctic and Nearctic subregions), Holotropical (Neotropical, Afrotropical, Oriental, and Australia-Tropical subregions), and Austral (Andean, Cape, Neozealandic, Neoguinean, Australian-Temperate, and Antarctic subregions).

Taxa selection: Distributional data of 154 taxa inhabiting the Andean subregion were analyzed. The included taxa (one fungus, 53 animals, 100 vascular plants) ranked from genus to family.

Construction of individual tracks: Distribution of 154 taxa was plotted on maps and their separate distribution areas connected with lines (individual tracks). The subregions or provinces inhabited by each taxon were connected according to their geographical proximity. Figure 5.5b is an example of an individual track belonging to the plant genus *Aristotelia.*

Construction of generalized tracks: Coinciding individual tracks resulted in the following generalized tracks: (I) Puna, Central Chilean, Subantarctic, and Patagonian provinces; (II) Central Chilean, Patagonian, and Subantarctic provinces; (III) Subantarctic and Patagonian provinces; (IV) Subantarctic province, Australian-Temperate, and Neozealandic subre-

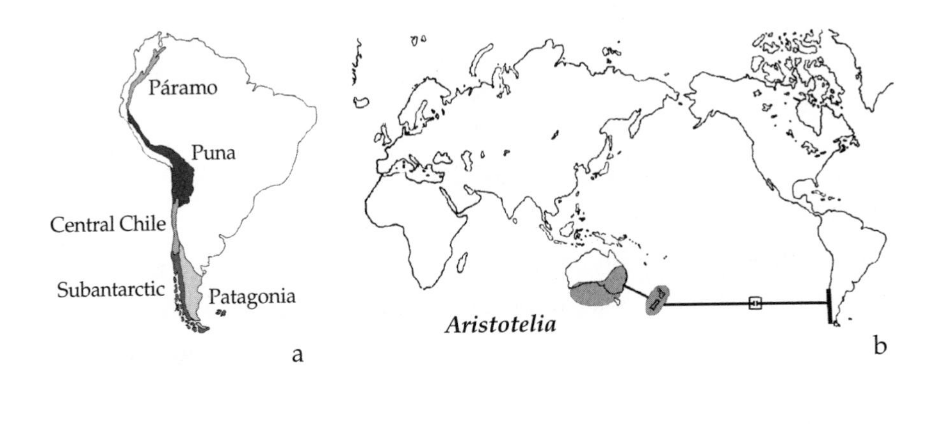

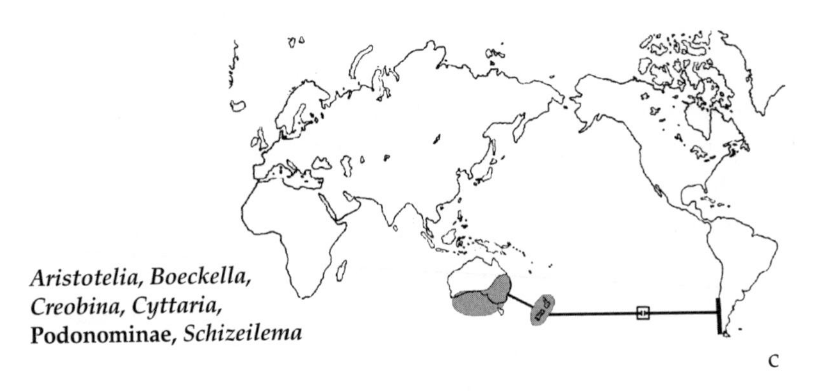

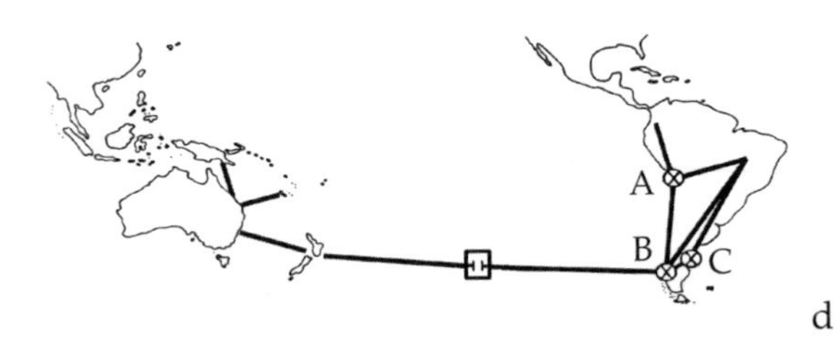

FIGURE 5.5. Panbiogeography of the Andean Subregion. *(a)* Andean Subregion provinces; *(b)* individual track of the genus *Aristotelia; (c)* generalized track represented by *Aristotelia, Boeckella, Creobina, Cyttaria,* Podonominae, and *Schizeilema; (d)* all generalized tracks and the three nodes found: A, B, and C.

gions; (V) Subantarctic province and Neozealandic subregion; (VI) Andean and Neotropical subregions; (VII) Páramo, Puna provinces, and Neotropical subregion; (VIII) Patagonian province and Neotropical subregion. Figure 5.5c shows the generalized track IV that connects the Subantarctic province, Australian-Temperate and Neozealandic subregions and is supported by six taxa.

Node delimitation: The intersection of the generalized tracks allows three nodes to be identified: Puna province (A), Subantarctic province (B), and Patagonian province (C) (Figure 5.5d).

The generalized tracks obtained establish three patterns: (1) Andean endemic pattern, represented by the generalized tracks I–III; (2) Austral pattern, represented by the generalized tracks IV–V; and (3) Tropical pattern, represented by the generalized tracks VI–VIII. The Austral and Tropical patterns would reveal that the biota of the Andean subregion has a complex or hybrid origin where two different ancestral biological and geological worlds met and combined. The Austral pattern would reflect the existence of an ancient Austral biota, with Gondwana events likely to have played a major role in its evolution, whereas the origin of the Neotropical pattern would be linked to a more recent history, especially the uplift of the Andes and further glaciation.

CASE STUDY 2: TRACK COMPATIBILITY OF TANDILIA AND VENTANIA BASED ON ASTERACEAE

Tandilia and Ventania are the only systems of mountain ranges situated in a grassy steppe or "pampas" in the political province of Buenos Aires, Argentina. From a geological point of view both areas do not appear to be closely related. Tandilia originated in the Proterozoic Age, being the oldest nucleus in Argentina. Ventania dates from the Paleozoic. Biologically, Tandilia and Ventania exhibit high taxon diversity and endemicity. Based on their particular biodiversity the ranges are considered to be "orographic islands." Asteraceae is the second most numerous plant family in genera and species in Ventania and Tandilia. A compatibility track analysis was performed to seek historical explanations for the high diversity and endemicity of Asteraceae in these areas (Crisci et al., 2001).

Delimitation of areas: Thirteen areas were delimited for this study on the basis of several overlapping taxon distributions and phytogeographical and geological criteria (Figure 5.6): Central Chile (CC), Central and North America (CNA), Chaco (CHA), Mahuidas (MA), Northern Andes (NA), Pampa (PAM), Patagonia (PAT), Sierras Pampeanas (SPA), Sierras Subandinas (SA), Southern Brazil (SB), Tandilia (TA), Uruguay (UR), and Ventania (VE).

Selection of taxa: Distributional data of the 112 species and infraspecific taxa of Asteraceae inhabiting Tandilia and Ventania were analyzed.

Construction of individual tracks: Distributions of 112 taxa were plotted on maps and their separate distribution localities were connected with lines (individual tracks) according to their geographic proximity.

Construction of generalized tracks: The compatibility track method was applied (Craw, 1988a). The constructed data matrix (13 areas × 112 tracks) was analyzed using SECANT 2.2 (Salisbury, 1999). The analysis of the data matrix resulted in four large cliques (generalized tracks) of 30 individual tracks each. The intersection of 27 individual tracks common to the four largest cliques is identified as the fifth clique or generalized track. In this way, six individual tracks are combined in groups of three with the 27 aforementioned individual tracks to form the four cliques of 30 individual tracks. The main distributional pattern obtained connects Tandilia and Ventania with Southern Brazil, Pampa, Uruguay, and Sierras Pampeanas (Fig. 5.6).

The main distributional pattern connecting Tandilia and Ventania with Southern Brazil, Pampa, Uruguay, and Sierras Pampeanas agrees with previous hypotheses (for example, Frenguelli, 1950; Ringuelet, 1956). Although Tandilia and Ventania are geographically close to each other, the geological evidence suggests that both resulted from independent geological processes at different geological times. The results reflect this fact, showing that Tandilia is more closely related to Uruguay and southern Brazil than to Ventania. The family Asteraceae probably originated in South America during the Oligocene (Stuessy et al., 1996). Two

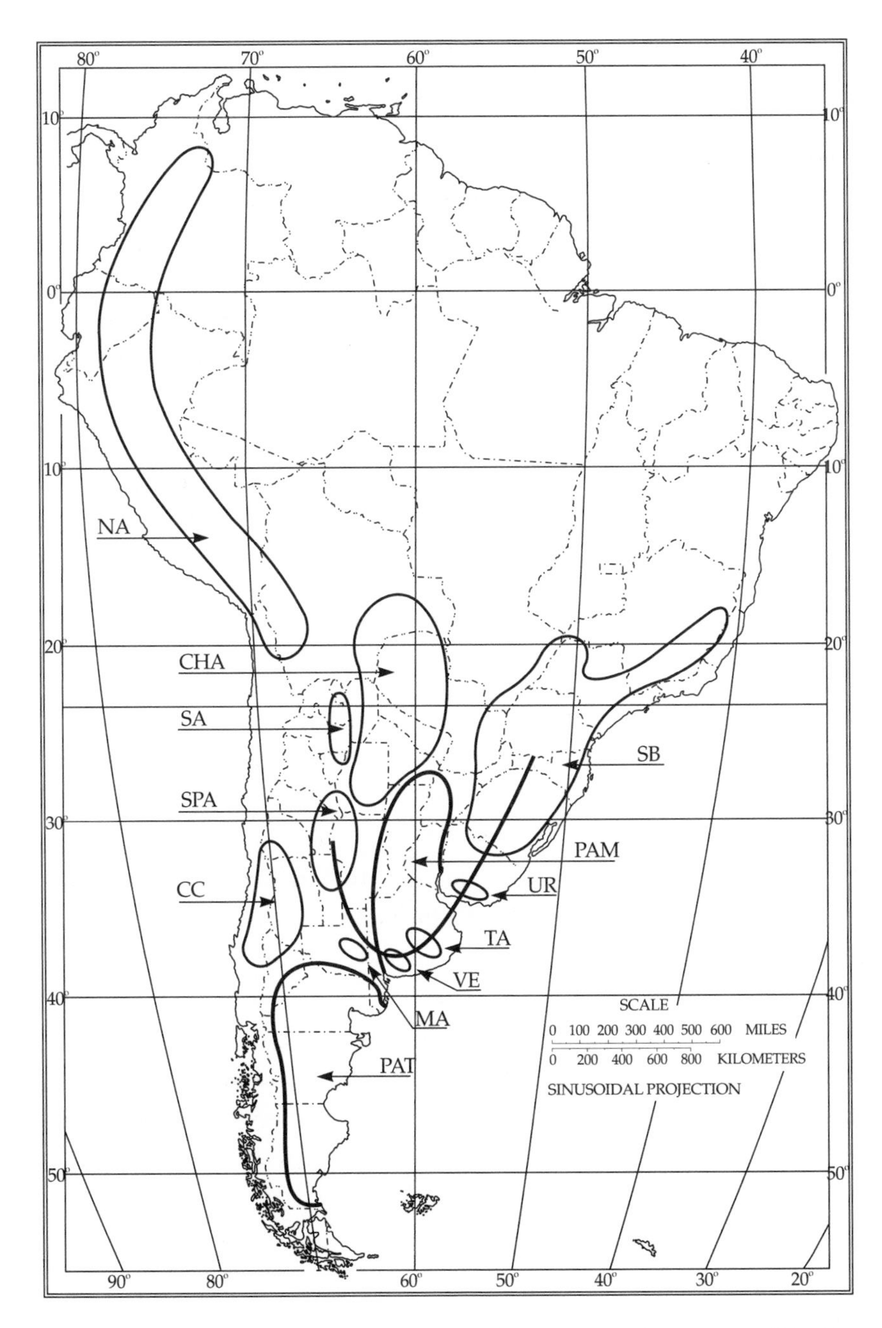

FIGURE 5.6. Panbiogeography of Tandilia and Ventania. Analyzed areas: CC = Central Chile, CHA = Chaco, MA = Mahuidas, NA = Northern Andes, PAM = Pampa, PAT = Patagonia, SPA = Sierras Pampeanas, SA = Sierras Subandinas, SB = Southern Brazil, TA = Tandilia, UR = Uruguay, and VE = Ventania. The main distributional pattern obtained is indicated as a generalized track represented by a line. Central and North America (CNA) is not shown.

major geological events could have affected the Asteraceae evolution, the uplift of the Andes and the Pleistocene glaciations. The former resulted in drier climatic conditions beginning in the Miocene (25.5 million years ago) and in additional uplift of several mountain systems. The Quaternary glaciations resulted in dry and wet cycles that caused fragmentation and differentiation of populations (Simpson Vuilleumier, 1971; B. B. Simpson, 1975). The main distributional pattern found in this study and the high endemicity of Asteraceae now found in Tandilia and Ventania could be explained as a consequence of these events, which resulted in eventual fragmentation of population in more elevated regions, restricted now to rocky and loose soils.

RESEARCH USING PANBIOGEOGRAPHY

Among the studies that use panbiogeography we recommend those by Craw on the Southern Hemisphere (Craw, 1985) and on New Zealand and the Chatham Islands (Craw, 1988a); by Heads on the Auckland Islands (Heads, 1986), New Zealand (Heads, 1990), the South Pacific area (Heads, 1999), and New Guinea (Heads, 2001); by Grehan on the Southern Hemisphere (Grehan, 1988b) and the Galapagos archipelago (Grehan, 2001b); by Climo (1989) on New Zealand; by Morrone (1992, 1993a, 1996b) on South America; Hadju (1995, 1998) on marine environments; Weston and Crisp (1996) on Trans-Pacific areas; Posadas and colleagues (1997) on South America; Aguilar-Aguilar and Contreras-Medina (2001) on marine environments; Contreras-Medina and Eliosa (2001) on America; and Franco (2001) on South America.

6

CLADISTIC BIOGEOGRAPHY

CLADISTIC BIOGEOGRAPHY was originally developed by Donn Rosen, Gareth Nelson, and Norman Platnick (Nelson, 1973, 1974; Rosen, 1976; Nelson & Platnick, 1981). Recently, Christopher Humphries and Lynne Parenti published a book providing an exposition of the history, methods, and applications of cladistic biogeography (Humphries & Parenti, 1999). This approach assumes that the correspondence between taxonomic relations and the relations among the areas is biogeographically informative. Basically, cladistic biogeography poses itself the following questions:

What are the world areas that house endemic taxa? How do we recognize them, how many are there, and where are they?

Given some number of areas of endemism, are the taxa inhabiting them phylogenetically related by some identifiable pattern to taxa elsewhere?

If there are one or more general patterns of relationships among areas; do those patterns show correlation with identifiable events in Earth history?

Cladistic biogeography bases itself essentially in the search for a pattern in the relations among areas of endemism that arises repeatedly in different taxon phylogenies, which may correspond to events in Earth history.

There is no doubt that cladistic biogeography is inspired by Hennig's phylogenetic systematics (1950) and Croizat's panbiogeography (1958), although it is important to note that Croizat emphatically rejected any conceptual relation between panbiogeography and cladistic biogeography (Croizat, 1982). Furthermore, it has been stated that the track concept is related to the area cladogram concept.

The application of cladistic biogeography supposes that the areas of endemism have been previously determined (see chapter 1). Then area cladograms must be constructed from different taxon cladograms, and likewise one or more general area cladograms (Fig. 6.1) (Crisci & Morrone, 1990, 1992b; Morrone & Crisci, 1995).

CONSTRUCTING THE AREA CLADOGRAMS

The area cladograms are constructed by replacing the names of the terminal taxa in the taxon cladograms with the areas of endemism that the terminal taxa inhabit. Obtaining the area cladograms is simple when each taxon is endemic to one area and every area is inhabited by only one taxon. Of course, it becomes more complicated when we find absent (= missing) areas, redundant distributions, and widespread taxa. Absent areas are those that are not inhabited by any taxa in the cladogram, so that that particular area will be absent from the area cladogram of that taxon. In a redundant distribution, at least one of the areas is inhabited by more than one taxon, so that in the area cladogram of that taxon some of them will be repeated. Widespread taxa occur when at least one taxon is distributed in more than one area, so that these areas will appear together in the corresponding area cladogram.

In these three cases the area cladograms must be changed to resolved area cladograms. To achieve this, procedures have been proposed, called assumptions 1 and 2 (Nelson & Platnick, 1981; Humphries & Parenti, 1986; Page, 1988, 1990; Nelson & Ladiges, 1991b; van Veller et al., 1999,

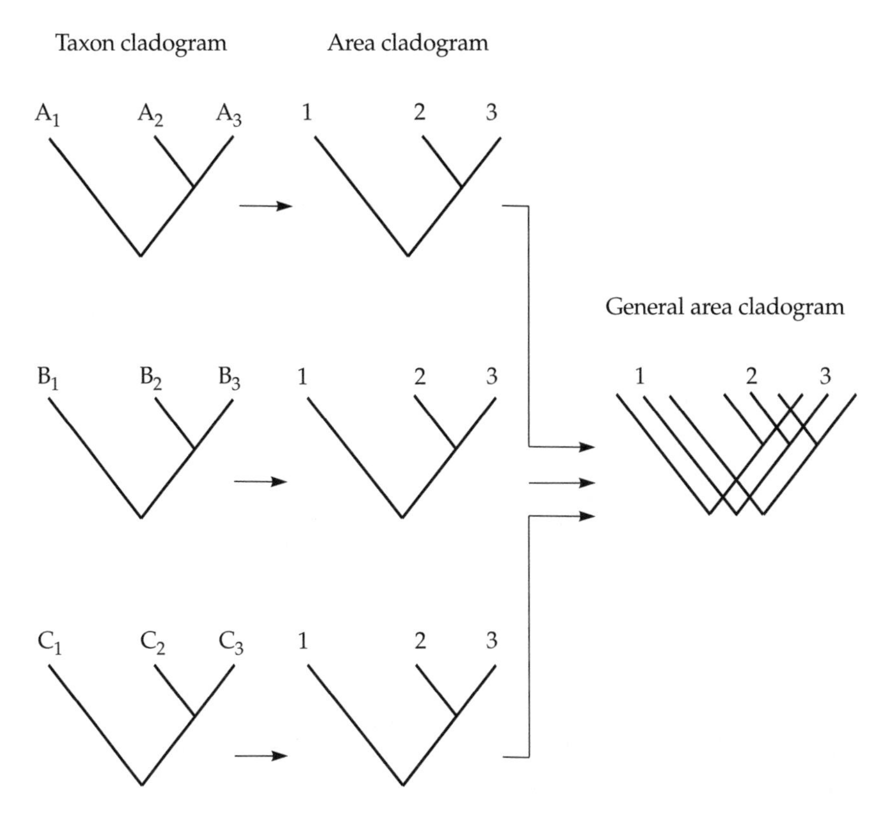

FIGURE 6.1. Steps in a cladistic biogeographic analysis. Construction of area cladograms from taxon cladograms and derivation of a general area cladogram. Taxa: A, B, C. Areas: 1, 2, 3.

2000) and assumption 0 (Wiley, 1987, 1988; Zandee & Roos, 1987; van Veller et al., 1999, 2000).

In Figure 6.2 we observe how each assumption resolves a three-taxa cladogram in the case of widespread taxa. Assumption 0 considers the two areas inhabited by the same taxon as a monophyletic group; assumption 1 considers them as a monophyletic or paraphyletic group; and assumption 2 allows all the possible locations within the two areas, so that the relations among the areas may be mono-, para-, or polyphyletic. In

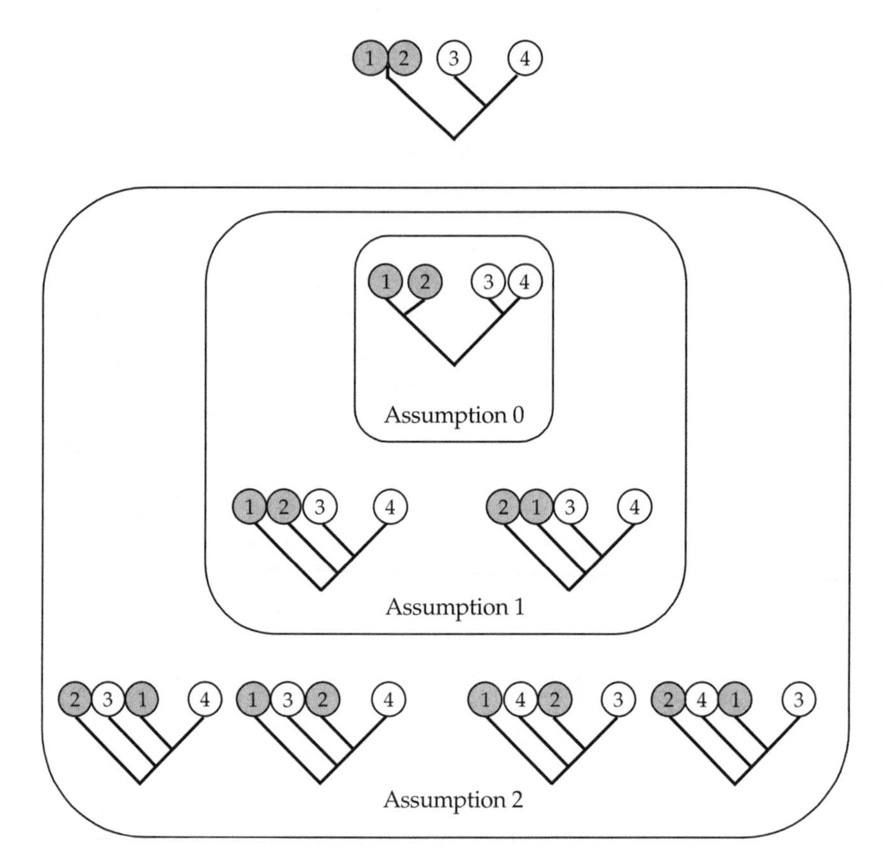

FIGURE 6.2. Area cladogram with a widespread taxon in areas 1 and 2, and derivation in resolved area cladograms under assumptions 0, 1, and 2.

the abovementioned figure we see how among the cladograms obtained under assumption 1 the cladogram obtained under assumption 0 is included, and that the cladograms obtained under assumption 2 include all the cladograms obtained under assumptions 0 and 1. Absent areas are considered not informative by assumptions 1 and 2, and as primitively absent under assumption 0. Concerning redundant distributions, under assumptions 0 and 1 if two taxa are present in the same area, their occurrences are equally valid, whereas under assumption 2 each occurrence of a redundant distribution is considered separately (for example, in different resolved area cladograms). The assumptions 0, 1, and 2 are not mutually exclusive, so that the treatment of each special case can be combined under a different assumption (for example, widespread taxa under assumption 2 and the redundant distributions and absent areas under assumption 0).

Some biogeographers generally prefer to use assumption 2. Yet, its implementation can produce many resolved area cladograms when the data set is complex. Nelson and Ladiges (1992) consider that most of the present implementations of assumption 2 are faulty because they may hide the real complexity. These authors suggest that the set of cladograms obtained under assumption 2 could be resolved afterward if nodes are assessed in terms of the three-area statement analysis, reducing the widespread taxa for the benefit of the endemic ones. A possible way of reducing the impact of redundant areas and widespread taxa would be to take them off before the analysis. Page (1994a) has found fault with the application of assumptions 1, 2, and 0 because he judges them to be mere algorithmic solutions without an optimizing criterion.

OBTAINING THE GENERAL AREA CLADOGRAMS

General area cladograms can then be obtained using area or resolved area cladograms. Many methods have been suggested to accomplish this procedure (see under "Taxonomy of Methods" in the Introduction), but here we will present only five: component analysis, Brooks parsimony analysis, the three-area statement, paralogy-free subtrees, and the integrative method.

Component Analysis

By employing this analysis, proposed by Nelson and Platnick (1981), general area cladograms are obtained from the intersection of the set of cladograms found through the application of assumptions 0, 1, and 2. If the resolved cladograms do not intersect, i.e., there is no resolved area cladogram common to all used taxa, then a consensus tree is constructed among all resolved cladograms obtained. If the intersection contains multiple general area cladograms, all of them are considered to be equally valid, because the history of the relations among the areas is not necessarily unique, as it is for the history of a taxon. In Figure 6.3 the application of assumption 2 produces 11 area cladograms for a taxon cladogram with one widespread taxon, two area cladograms for a taxon cladogram with redundant distribution, and seven area cladograms for a taxon cladogram with one missing area. The intersection produces two general area cladograms (indicated with squares). This method can be performed by applying COMPONENT 1.5 (Page, 1989).

Brooks Parsimony Analysis (BPA)

This method, proposed by E. O. Wiley (1987), is based on the ideas developed by D. R. Brooks (1985, 1990) for historical ecology studies. Brooks's method implicitly contains the idea of assumption 0, even though the formal enunciation of this assumption was developed later. So, BPA applies a principle similar to assumption 0, the difference being only that it considers the absent areas as uninformative instead of primitively absent. BPA consists of two analytical steps, primary BPA and secondary BPA (see van Veller & Brooks, 2001). Primary BPA involves constructing an area × cladogram components matrix (by cladogram components we mean taxa and monophyletic groups, that is, the terminal cladogram branches and each of its nodes). An additional row has to be added to the data matrix, which is typically given the name "ancestor" (Lieberman, 2000), acting as an outgroup. This additional row is coded with all entries equal to 0. The matrix is then analyzed using a maximum parsimony algorithm and results in one or more general area cladograms. The application of maximum parsimony algorithms may be carried out with such software as Hennig86 (Farris, 1988), PAUP (Swofford, 2000), PHYLIP (Felsenstein, 1993), or NONA (Goloboff, 1996).

Figure 6.4 shows an example of the application of BPA. In Figure 6.4a–d four area cladograms of different taxa with their respective codification are shown:

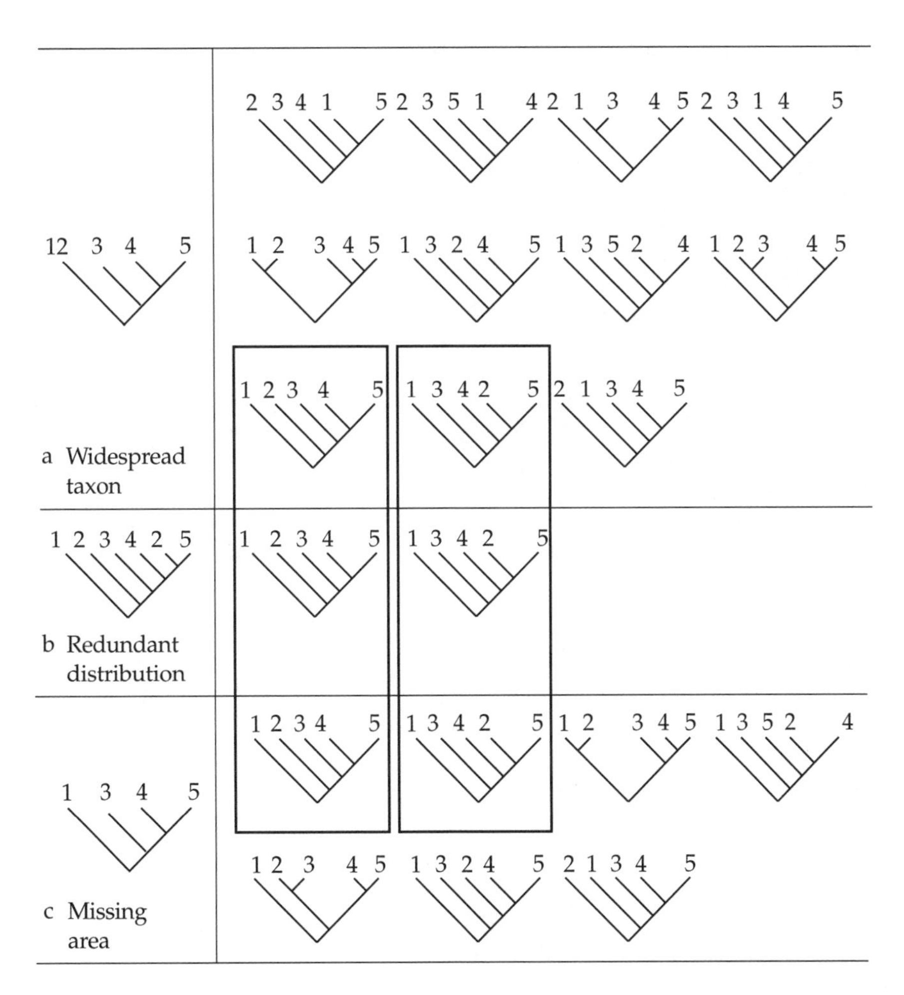

FIGURE 6.3. Application of component analysis using assumption 2. *(a)* Area cladogram including a widespread taxon; *(b)* area cladogram including a redundant distribution; *(c)* area cladogram including a missing area. Intersection of the three sets of resolved area cladograms (marked with squares) represents the general area cladograms. Areas: 1–5.

(a) Cladogram of one taxon with complete data and without ambiguity.

(b) Cladogram with an absent area, that is, inhabited by no taxon of the group, that in this technique is considered as uninformative (codified as "?" in data matrix).

(c) Cladogram with a widespread taxon (present in areas 1 and 2).

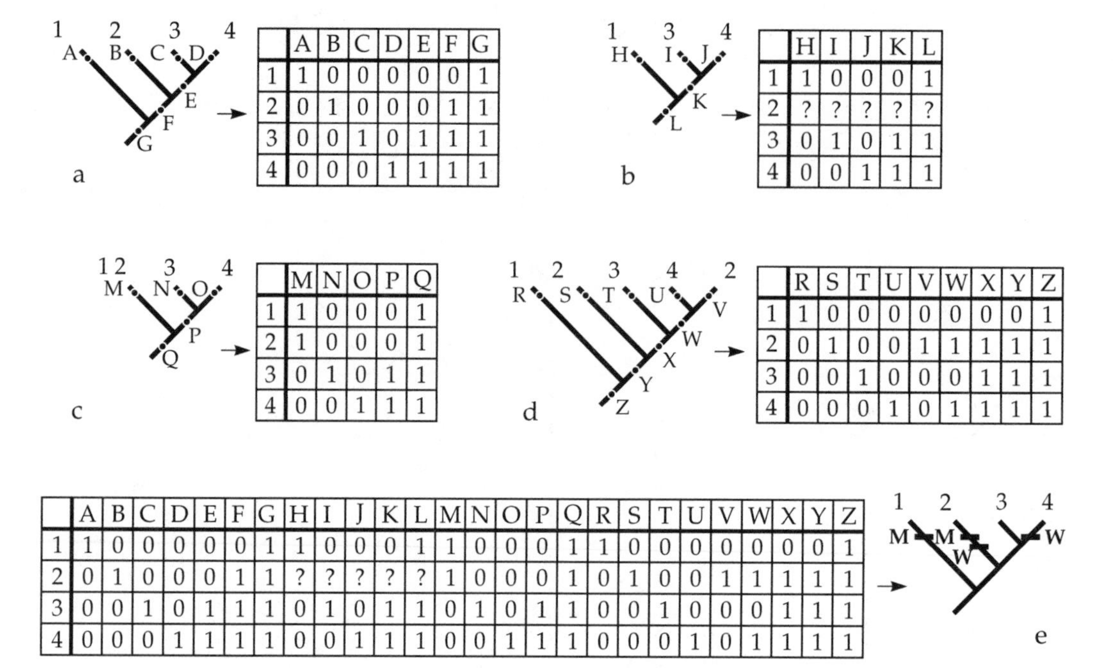

	A	B	C	D	E	F	G
1	1	0	0	0	0	0	1
2	0	1	0	0	0	1	1
3	0	0	1	0	1	1	1
4	0	0	0	1	1	1	1

	H	I	J	K	L
1	1	0	0	0	1
2	?	?	?	?	?
3	0	1	0	1	1
4	0	0	1	1	1

	M	N	O	P	Q
1	1	0	0	0	1
2	1	0	0	0	1
3	0	1	0	1	1
4	0	0	1	1	1

	R	S	T	U	V	W	X	Y	Z
1	1	0	0	0	0	0	0	0	1
2	0	1	0	0	1	1	1	1	1
3	0	0	1	0	0	0	1	1	1
4	0	0	0	1	0	1	1	1	1

	A	B	C	D	E	F	G	H	I	J	K	L	M	N	O	P	Q	R	S	T	U	V	W	X	Y	Z
1	1	0	0	0	0	0	1	1	0	0	0	1	1	0	0	0	1	1	0	0	0	0	0	0	0	1
2	0	1	0	0	0	1	1	?	?	?	?	?	1	0	0	0	1	0	1	0	0	1	1	1	1	1
3	0	0	1	0	1	1	1	0	1	0	1	1	0	1	0	1	1	0	0	1	0	0	0	1	1	1
4	0	0	0	1	1	1	1	0	0	1	1	1	0	0	1	1	1	0	0	0	1	0	1	1	1	1

FIGURE 6.4a–e. Primary Brooks parsimony analysis. *(a)* Area cladogram with complete data and no ambiguity and its derived matrix ("outgroup" is not shown); *(b)* area cladogram including a missing area and its derived matrix; *(c)* area cladogram including a widespread taxon and its derived matrix; *(d)* area cladogram including a redundant distribution and its derived matrix; *(e)* data matrix (areas × components) and resulting general area cladogram obtained after the application of a maximum parsimony algorithm, showing the DELTRAN optimization of the homoplastic characters.

(d) Cladogram with a redundant distribution (two taxa inhabiting area 2).

All information is combined in a unique matrix that, after applying a maximum parsimony algorithm, results in a general area cladogram, which is illustrated showing the optimization of the homoplastic characters using the DELTRAN option, which maximizes parallel changes and minimizes reversions (Fig. 6.4e).

Primary BPA is based on the null hypothesis of vicariant speciation (simple vicariance) and, therefore, it assumes single unique histories for all areas (each area has a single history with respect to all taxa begin analyzed). According to Brooks and colleagues (2001), episodes of extinction (reversions to 0 in primary BPA general area cladograms) do not falsify the null hypothesis of simple vicariance, as the absence of evidence can neither corroborate nor falsify a hypothesis. True falsification of the null hypothesis of simple vicariance comes in the form of parallelisms (generated by redundant distributions and widespread taxa) in the primary BPA general area cladogram. Primary BPA can indicate whether or not the data support a general pattern of area relationship, and whether or not there are exceptions (falsifiers) to that pattern. But it does not permit us to represent those falsifications in an effective manner. Brooks (1990) and Brooks and McLennan (1991) resolved this methodological and conceptual problem by proposing that we represent each instant of falsification with an extrarepresentation of the area having the reticulate history. Van Veller and Brooks (2001) called this protocol "secondary BPA." Secondary BPA consists of the construction of a new area × cladogram components matrix, in which the parallelisms are solved via area duplication, and by reapplying a maximum parsimony algorithm generating a secondary BPA general area cladogram. In our example (Fig. 6.4a-e), the primary BPA general area cladogram shows four parallelisms, all of them involving area 2. Treating area 2 as three separate areas (2a, 2b, and 2c) solves via extrarepresentation the reticulate history of area 2. Area 2a represents the original congruent relationships of area 2 with the general pattern. Areas 2b and 2c represent the incongruent relationships of area 2 with the general pattern. The new matrix (Fig. 6.4f) of 6 areas × 26 components is analysed applying a maximum parsimony algorithm resulting in a single secondary BPA area cladogram (Fig. 6.4g). Representing area 2 three times integrates the incongruent elements with the general pattern by choosing the result (secondary BPA general area cladogram), which postulates the fewest number of area duplications, each one of which depicts a falsification of the null hypothesis of simple vicariance (for a detailed discussion of the methodology, see Brooks et al., 2001).

Lieberman and Eldredge (1996) proposed a modified form of BPA.

	A	B	C	D	E	F	G	H	I	J	K	L	M	N	O	P	Q	R	S	T	U	V	W	X	Y	Z
1	1	0	0	0	0	0	1	1	0	0	0	1	1	0	0	0	1	1	0	0	0	0	0	0	0	1
2a	0	1	0	0	0	1	1	?	?	?	?	?	0	0	0	0	1	0	1	0	0	0	0	0	1	1
2b	?	?	?	?	?	?	?	?	?	?	?	?	1	0	0	0	1	?	?	?	?	?	?	?	?	?
2c	?	?	?	?	?	?	?	?	?	?	?	?	?	?	?	?	?	0	0	0	0	1	1	1	1	1
3	0	0	1	0	1	1	1	0	1	0	1	1	0	1	0	1	1	0	0	1	0	0	0	1	1	1
4	0	0	0	1	1	1	1	0	0	1	1	1	0	0	1	1	1	0	0	0	1	0	1	1	1	1

f

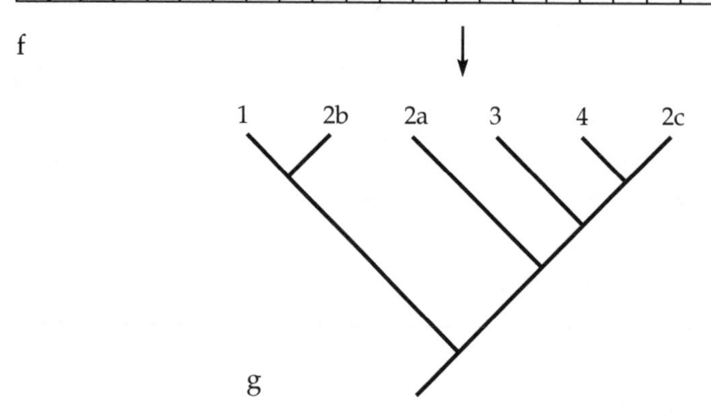

FIGURE 6.4f–g. Secondary Brooks parsimony analysis. *(f)* Matrix of area × cladogram component, with area 2 treated as three separate areas; *(g)* general area cladogram for areas 1–4 produced by secondary BPA. Note that representing area 2 three times integrates the reticulate history of this area with the general pattern. Areas: 1–4; components: A–Z.

According to these authors the way in which a BPA data matrix is coded permits one to recover only the vicariance signal of the data. To solve this problem they proposed a different coding procedure using Fitch optimization (Fitch, 1971) to determine the distribution at each node. They argued that there are three different things that can happen to the geographic range between adjacent ancestral and descendant nodes and between ancestral nodes and their corresponding terminal taxa—the geographic range could contract, expand, or remain constant. A contraction of the geographic range is likely to be related to a vicariant event; conversely, an expansion in it could be related to geodispersal (see "Space-Time Processes" in the Introduction). When the range remains constant it implies no vicariance and no geodispersal. Lieberman and Eldredge proposed the creation of two separate data matrices, one for vicariance and one for geodispersal (for details of this protocol, see Lieberman and Eldredge, 1996). This variant of BPA was mentioned neither by Brooks and colleagues (2001) nor by van Veller and Brooks (2001) in their two updated papers on BPA.

BPA has been criticized by Page (1990, 1994a, 1994b) and Ronquist and Nylin (1990). One common criticism is that primary BPA tends to overestimate dispersal events due to the fact that the binary codes derived from an area cladogram are not independent.

Three-Area Statements (TAS)

This method (Nelson & Ladiges, 1991a,b) has its origin in the three-item statement analysis (Nelson & Platnick, 1991), which is intended to resolve the relationships among three taxa. It suggests that given three taxa, two of them will be more strongly related to each other than to the third. This analysis transforms the character distribution in all possible three-taxon statements to be an expression of the relationships among these three taxa. Any systematic data set for more than three taxa can be transformed into three-taxon statements, which when analyzed can produce more coherent results than those obtained directly from the original data. In the TAS method the distribution data of every area cladogram are codified as three-area statements. The result is an area × three-area statements data matrix. This data matrix can be obtained with the TAS program (Nelson & Ladiges, 1992) implemented for assumptions 0 and 1. Assumption 2 can be applied by prior manipulation of the data set or with the TASS program (Nelson & Ladiges, 1995). The matrix produced with TAS is analyzed using a maximum parsimony algorithm. Figure 6.5a–e shows TAS applied in the same example used for BPA, with the matrices corresponding to the three-area statements and the resulting general area cladogram.

The taxonomic application of the TAS method has been criticized (see for example Kluge, 1993).

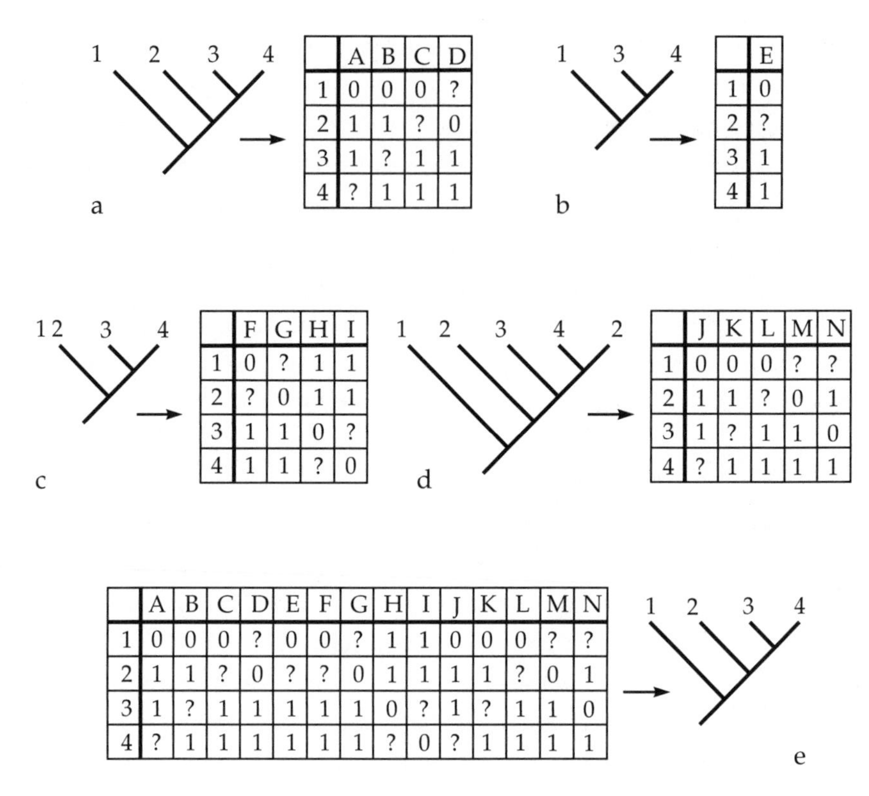

	A	B	C	D
1	0	0	0	?
2	1	1	?	0
3	1	?	1	1
4	?	1	1	1

	E
1	0
2	?
3	1
4	1

	F	G	H	I
1	0	?	1	1
2	?	0	1	1
3	1	1	0	?
4	1	1	?	0

	J	K	L	M	N
1	0	0	0	?	?
2	1	1	?	0	1
3	1	?	1	1	0
4	?	1	1	1	1

	A	B	C	D	E	F	G	H	I	J	K	L	M	N
1	0	0	0	?	0	0	?	1	1	0	0	0	?	?
2	1	1	?	0	?	?	0	1	1	1	1	?	0	1
3	1	?	1	1	1	1	1	0	?	1	?	1	1	0
4	?	1	1	1	1	1	1	?	0	?	1	1	1	1

FIGURE 6.5. Three-area statements applied on the same example as figure 6.4. *(a)* Area cladogram with complete data and no ambiguity and its derived matrix; *(b)* area cladogram including a missing area and its derived matrix; *(c)* area cladogram including a widespread taxon and its derived matrix; *(d)* area cladogram including a redundant distribution and its derived matrix; *(e)* data matrix (areas × three-area statements) and resulting general area cladogram obtained after the application of a maximum parsimony algorithm. Areas: 1–4; three-area statements: A–N.

Paralogy-Free Subtrees

Geographic paralogy is made evident by the repetition of geographic areas in an area cladogram, which thereby presents redundant distributions. The so-called redundant or paralogous nodes then appear within

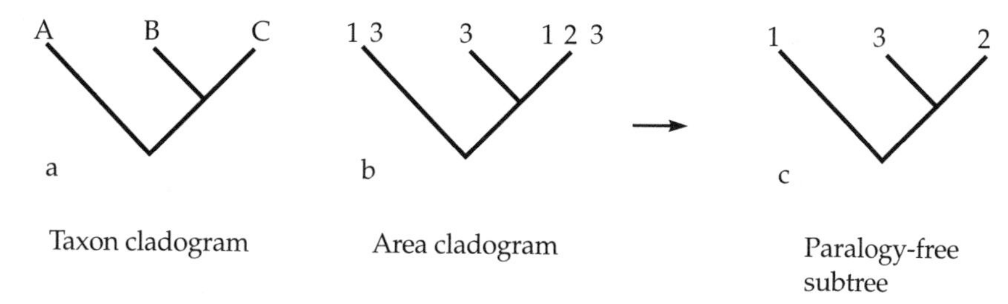

FIGURE 6.6. Paralogy-free subtrees. *(a)* Taxon cladogram of species A, B, and C; *(b)* corresponding area cladogram; 1, 2, and 3 represent the areas inhabited by taxa A, B, and C; the area cladogram shows paralogy due to area 3 is inhabited by the three taxa, and area 1 is inhabited by the taxa A and C; *(c)* paralogy-free subtree.

these cladograms, that is, those that contain at least two descendants whose areas are superposed (Page, 1988). These nodes cause ambiguity in the relationships among the areas. For example, if a three-taxa cladogram (A(B,C)) whose area cladogram could be stated as (1 3 (3, 1 2 3)), where 1, 2, and 3 are the areas inhabited by taxa A, B, and C, and this area cladogram is analyzed with assumption 2, it results in a sole informative statement of relation among the areas: (1 (2,3)) as shown in Figure 6.6. Nelson and Ladiges (1996) suggest a methodological solution to solve the paralogy problem, which they called paralogy-free subtrees. This method is applied by means of the TASS algorithm, which takes all unambiguous information from the trees (that is, identifies the paralogy-free subtrees of the area cladograms) and codifies the informative nodes as characters in a matrix.

Figure 6.7a–b shows two area cladograms that exhibit paralogy, the corresponding paralogy-free subtrees, and the area × components resulting matrices, obtained with the TASS program (Nelson & Ladiges, 1995). Both matrices can be combined into one, which is analyzed with a maximum parsimony algorithm to obtain the general area cladogram (Fig. 6.7c). There are two ways of carrying out the coding of the paralogy-free subtrees. The first consists of expressing the matrix as area × components

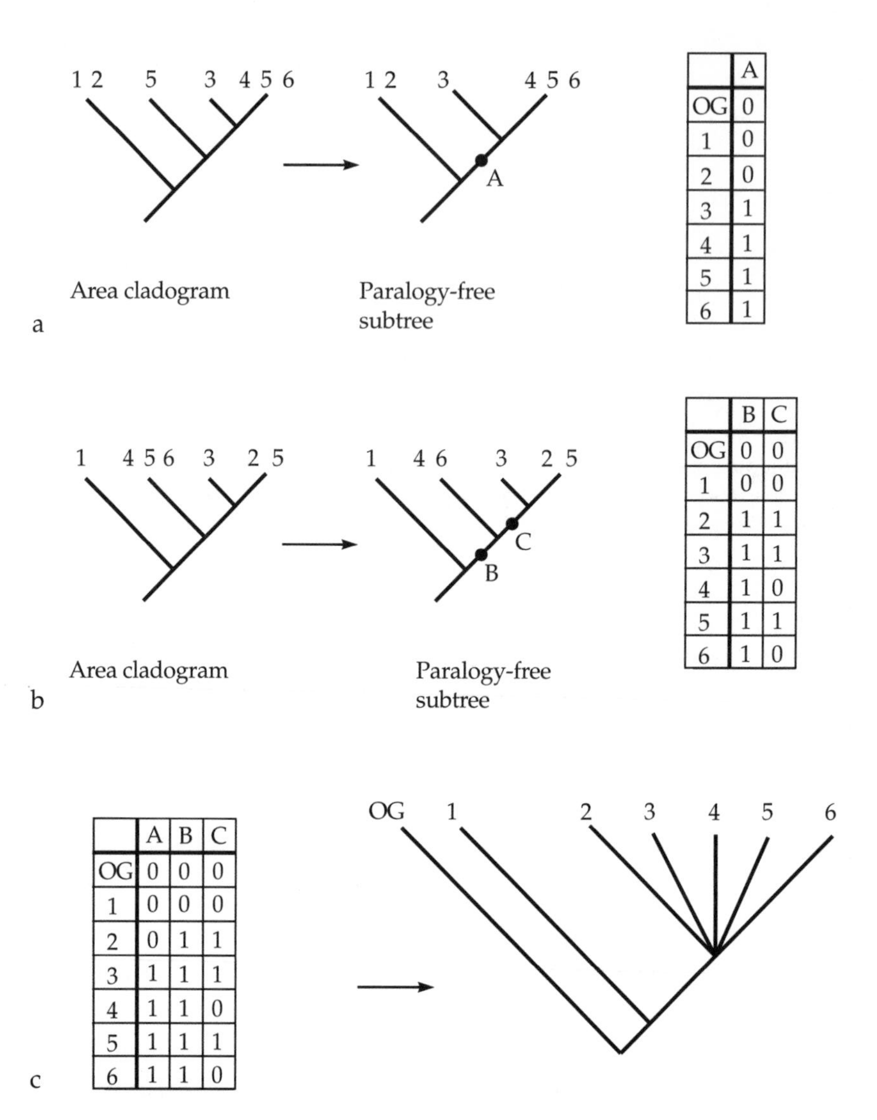

	A
OG	0
1	0
2	0
3	1
4	1
5	1
6	1

	B	C
OG	0	0
1	0	0
2	1	1
3	1	1
4	1	0
5	1	1
6	1	0

	A	B	C
OG	0	0	0
1	0	0	0
2	0	1	1
3	1	1	1
4	1	1	0
5	1	1	1
6	1	1	0

FIGURE 6.7. Paralogy-free subtrees. *(a–b)* Area cladograms of two taxa showing their corresponding paralogy-free subtrees and the derived data matrix (areas × components) for each one, applying TASS; *(c)* data matrix (areas × components) resulting from the combination of a and b matrices, and general area cladogram obtained after the application of a maximum parsimony algorithm. Areas: 1–6; components: A–C; OG: hypothetical area with all absences as ancestral states.

of the paralogy-free subtrees. This matrix is directly constructed by the TASS program (see Fig. 6.7a–c). The second consists of expressing the matrix as areas × three-area statement, which arose from the paralogy-free subtrees. To construct this matrix it is necessary to input the area matrix obtained by TASS in the TAS program, which makes the conversion. In both cases the matrices contain a hypothetical external area coded 0 for all the components. The matrix is then analyzed using a maximum parsimony algorithm to obtain one or more general area cladograms.

Integrative Method

The integrative method (Morrone & Crisci, 1995) consists of using different methods in different steps of a biogeographic analysis, restricting their uses to specific problems. A historical biogeographic analysis should include at least three steps: recognition of spatial homology, identification of areas of endemism, and formulation of area relationship hypotheses.

The first step consists of determining if the taxa analyzed belong to the same biota. A panbiogeographic procedure could be employed to find generalized tracks, which represent ancestral biotas and spatial homologies (see chapter 5). Each generalized track then should be analyzed separately, thus avoiding the extreme incongruent patterns that result from mixing different ancestral biotas in the same analysis.

Once biogeographic homologies have been recognized, areas of endemism must be identified (see chapter 1). To do so, the parsimony analysis of endemicity methodology based in quadrats can be applied (see chapter 7).

When the areas of endemism have been identified and used to convert the taxon cladogram in area cladograms, a hypothesis can be formulated about the relationships among the areas. Page and Lydeard (1994) have suggested three criteria to select the taxa for this step: To maximize the endemicity, make an exhaustive sampling in each clade, and include as many relevant areas as possible. Once the area cladograms of the selected taxa are obtained, any of the cladistic biogeographic methods for obtaining general area cladograms can be applied.

COMPARING THE CLADISTIC BIOGEOGRAPHIC METHODS

Morrone and Carpenter (1994) compared the application of different methods of cladistic biogeography (component analysis, primary BPA, TAS). They reconciled the trees of the different data sets superposing the area cladograms onto the general area cladograms obtained from each one of these three methods and calculated the error items (that is, the quantity of nodes that must be added to the general area cladogram to be able to explain all the original area cladograms).

They also applied two additional criteria, the quantity of general area cladograms obtained by each method and their resolution grades (optimizing the minor number of general area cladograms and their degree of resolution). The result of this procedure was that none of the methods was demonstrated to be consistently better than the others, a result that could be influenced by the different sources of ambiguity for each method, which seem to affect each one differently. For Morrone and Carpenter (1994) the principal sources of ambiguity are the dispersal and the speciation events independent from the area fragmentation, which combine with the extinction and the sampling errors in the taxa distribution. Thus, for instance, the primary BPA is affected by dispersals more than the component analysis, whereas the latter is more affected by the redundant distributions that result from sympatric speciation.

Robin Craw (1982, 1983) and Dan Polhemus (1996) have criticized the application of cladistic biogeography because of the relevance that this approach gives to biota fragmentation (vicariance) to the detriment of area hybridization (composite areas). Platnick and Nelson (1984) have rejected these criticisms, maintaining that cladistic biogeography investigates the fragmentation of biotas instead of assuming them.

CASE STUDY: CLADISTIC BIOGEOGRAPHY OF SOUTHERN SOUTH AMERICA

For more than a century biologists have suggested many hypotheses to explain southern South America biota origin and relationships with other temperate area biotas, such as Australia, New Zealand, and South Africa. Some authors have stated that South America is a hybrid continent and it should be divided biogeographically into two areas, one southern and temperate and one northern and tropical (Humphries, 1981; Patterson, 1981). With the goal of stating a hypothesis on the relationships among the southern part of South America and other areas, a cladistic biogeographic analysis was performed (Crisci et al., 1991a).

Taxon cladograms: Seventeen taxon cladograms were used, including a fungus (*Cyttaria*), plants (*Crinodendron-Dubouzetia-Peripentadenia*, Embothriinae, *Nothofagus-Fagus, Negria-Depanthus, Oreomyrrhis, Drapetes, Drimys, Aristotelia*), and animals (*Oxelytrum-Ptomaphila*, Diamesinae, Podonominae, Pseudopsinae, Metallicina, Siphlonuridae, *Eriococcus-Madarococcus, Nannochoristina*).

Area cladograms: In the aforementioned cladograms, the terminal taxa were replaced by the areas of endemism that they inhabit, and the corresponding area cladograms were obtained.

General area cladograms: Two different techniques were applied: primary BPA, applying assumption 0; and the component analysis, applying assumptions 1 and 2. These methods resulted in seven general area cladograms. Two were from primary BPA (Fig. 6.8a–b), and five were from component analysis (two under assumption 1 [Fig. 6.8c–d] and four under assumption 2 [Fig. 6.8d–g], one of which [Fig. 6.8d] was coincident with one of the two that were obtained by applying assumption 1).

Even if a unique hypothesis of the relations among the areas did not result from the general area cladograms, in all of the general area cladograms a repeated pattern can be observed. According to this pattern, the south and the north of South America never constitute a monophyletic group. Southern South America is related to the other austral areas (excepting South Africa), while northern South America is the sister area of North America or both form a trichotomy with South Africa. These results support the hypothesis of a hybrid origin for South American biota. Likewise, some of the general area cladograms show different relationships among southern South America and other southern areas. This

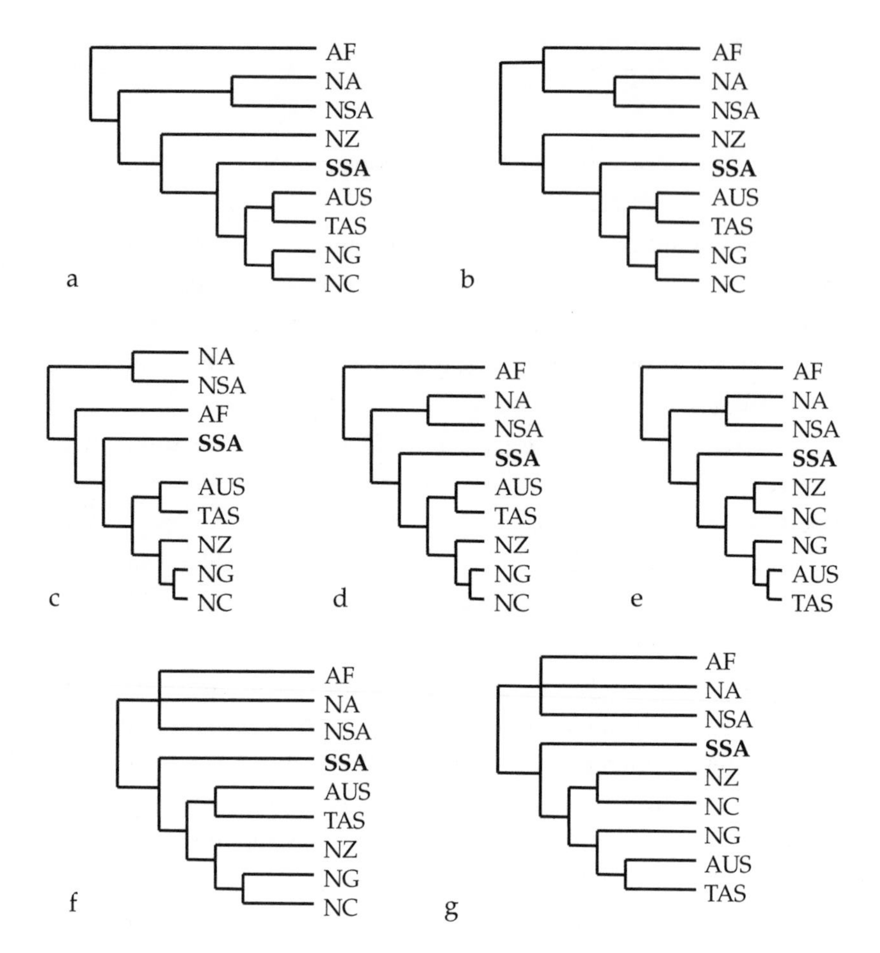

FIGURE 6.8. Some of the general area cladograms obtained in Crisci and colleagues' paper (1991a) showing the different relations of southern South America, with the other areas analyzed. *(a–b)* Cladograms obtained by primary Brooks parsimony analysis; *(c–g)* cladograms obtained by component analysis; *(c)* under assumption 1; *(d)* under assumptions 1 and 2; *(e–g)* under assumption 2. SSA = southern South America; NSA = northern South America; AUS = Australia; NG = New Guinea; NC = New Caledonia; TAS = Tasmania; NZ = New Zealand; AF = Africa; NA = North America.

would suggest that this area constitutes in itself a hybrid or composite area.

RESEARCH USING METHODS IN CLADISTIC BIOGEOGRAPHY

Many have been the empirical works in which cladistic biogeography has been applied, for example: Crisci and colleagues (1991b) on southern South America; Liebherr (1991, 1994a,b) on Mexico, Mexico and Central America, and North and Central America, respectively; Ladiges and colleagues (1992) on the Australian region; Morrone (1993b) on South America; Hadju (1995) on marine environments; Linder and Crisp (1995) on the Southern Hemisphere; De Meyer (1996) on the Hawaii Islands; Morrone and colleagues (1997) on Chile; Morrone and Urtubey (1997) on South America; Domínguez (1999) on America; Marshall and Liebherr (2000) on montane biota of Central America; Palmer and Cambefort (2000) on the Gibraltar Strait area; Brooks and McLennan (2001) on North America; and Brown and colleagues (2001) on the Australian region.

7

PARSIMONY ANALYSIS OF ENDEMICITY

PARSIMONY ANALYSIS of endemicity or PAE is a tool of historical biogeography for the investigation of the natural patterns of organism distribution (Rosen, 1988; Rosen & Smith, 1988). PAE classifies localities, quadrats, or areas (analogous to taxa, if compared with the analysis of phylogenetic systematics) according to their shared taxa (analogous to characters) by means of the most parsimonious solution (parsimony criterion), resulting in a hierarchical classification of the geographic units. Steven McLaughlin (1992) has demonstrated that floristic areas are actually arranged in a natural hierarchy. Brian Rosen (1988) originally proposed PAE methodology with the purpose of studying how fossils can provide spatial information in a geographic scale. This author uses localities as study units and that is why in this book we call his method "parsimony analysis of endemicity based on localities," to differentiate it from other PAE variations. Afterward, two other authors took Rosen's idea to analyze distributional data by means of the application of parsimony algorithms. The first was Robin Craw (1988a), whose objective was to investigate up to what level the geographic distribution of living organisms contains hierarchical information about the relationships of the areas such organisms inhabit. This author used areas of endemism previously defined on the basis of empirical data as study units. In this book we will call his method "parsimony analysis of endemicity based on areas of endemism." Finally, Juan Morrone (1994a) proposed the application of a maximum parsimony algorithm to delimit areas of endemism; to do it he used quadrats as analysis units used jointly by the existing shared taxa. We will call his method "parsimony analysis of endemicity based on quadrats"—a method mentioned in chapter 1.

PAE BASED ON LOCALITIES

Brian Rosen (1988) states that biotic regions or areas of endemism are difficult to establish, as there generally exist taxa that pass beyond the defined limits. Thus, he proposes to work with punctual localities, which represent a distribution sampling of every taxon. It is important to emphasize that endemicity respecting a locality is relative, that is, a taxon which is present in two localities and in no other is endemic to both; but at the same time it is not endemic to either of them, particularly when they are considered individually. Rosen (1988) postulates an analogy with the phylogenetic systematics from which the localities are interpreted as taxa. He specifies that each locality must be considered as if it was a specimen representing a taxon. In this way, previous assumptions related to area extension are avoided. According to Rosen (1988), PAE based on localities facilitates getting area cladograms of the localities. In these cladograms the localities are grouped by the presence of geographic synapomorphies (shared taxa between more than one locality) directly from the geographic distributions of the organism. The localities represent the study units. To construct such a matrix the presence of a taxon in a locality is coded as 1 and its absence as 0. The data are analyzed by means of a maximum parsimony algorithm (for example, with PAUP or Hennig86 programs). To facilitate the biogeographic analysis, Rosen proposes to modify the matrix of original data by considering the following:

> Any taxon common to all the localities or present in only one of them must be eliminated from the analysis as it is not informative

(it does not permit establishing groupings among the localities on the basis of shared presences);

the localities that show scarce presences compared with the other analyzed localities must be eliminated because low diversity is interpreted in the analysis as primitive; and

the localities that present exactly the same taxa must be considered as a unique analysis unit.

Usually, a hypothetical locality to root the tree with all the taxa absent is added to the data matrix. Rosen and Smith (1988) suggest analyzing the data without using this hypothetical locality and producing an unrooted tree. Once the tree is constructed, it is connected to the hypothetical locality through the real locality, the one that exhibits more absences. In phylogenetic systematics this technique of cladogram rooting is called Lundberg rooting (Lundberg, 1972).

In PAE, as originally proposed in paleontology, the data for a determined analysis must proceed from the same geological horizon (temporal equivalence). The cladograms obtained from different data sets collected in successive strata or geological horizons allow the researcher to establish if an ancient interchange event in a horizon is confirmed by more recent events in the next horizon.

PAE BASED ON AREAS OF ENDEMISM

Robin Craw (1988a) presents a variant of Rosen's method (1988) using areas that he terms high and low endemism as study units, in place of the punctual localities used in Rosen's method. Furthermore, Craw proposes to modify the original method, adding information on monophyletic clades. He then constructs data matrices of areas of endemism × taxa. To introduce the monophyletic clade information he codes in the following way:

Absences are always considered as 0;

the presence of a monophyletic taxon (for example, one species

within one genus with several species or a pair of sister species that represent by themselves a genus) is coded 1; and
the presence of a sister group of any of the taxa whose presence was coded as 1 is coded as 2, and so on.

He analyzes the data by applying a maximum parsimony algorithm and considering the multistate characters as unordered. A multistate character is that which has more than two states. Considering it as unordered means that any change from one state to another has the cost of one step. For Craw (1988a) character reversions are biogeographically interpreted as extinctions, whereas parallelisms are interpreted as dispersals.

Joel Cracraft (1991) proposes a method similar to Craw's (1988a). In Cracraft's method a data matrix of areas of endemism × taxa is constructed; the presence of a taxon in an area is coded as 1 and its absence as 0. To add the information related to larger taxonomic categories he adds a new column to the data matrix. For instance, to code a genus, he adds a new column of taxon to the matrix and in that column he assigns 1 to each area that is inhabited at least by one species of the genus. The difference between the coding of taxa of superior rank (for example, monophyletic taxa) applied by Craw and the one applied by Cracraft is different only in appearance since the matrices would be coincident if Craw's multistate characters are recoded as binary characters (0 or 1).

Cracraft emphasizes that his method differs from Rosen's (1988) in two ways. First, Cracraft's study uses predefined areas of endemism, whereas Rosen's (1988) used point-sample localities. Cracraft states that in this way some degree of historical and biogeographic nonrandomness is already assumed based on the original empirical distributions. Second, Rosen coded only one taxonomic level, whereas Cracraft's study imposes a hierarchical structure on the raw data by coding larger taxonomic categories when possible.

Cracraft states that the best way to create a root for the tree is to add a hypothetical external area in which all taxa are absent, but he does not mention Lundberg rooting. The data are analyzed applying a conventional program for cladistic analysis, which uses a maximum parsimony algorithm. The objective of this method is to obtain an area cladogram that shows the relationships among the areas of endemism used as units.

PAE BASED ON QUADRATS

Morrone (1994a) proposed to analyze the distributional information of taxa applying a maximum parsimony algorithm in a different way. The objective in this case is to delimit areas of endemism. The PAE based on quadrats comprises five steps (Posadas & Miranda-Esquivel, 1999):

Grid construction: The study area is divided into quadrats. The quadrats are not necessarily equal in shape and size, so the grid can be adjusted to the resolution expected in a particular subarea (Fig. 7.1a). Some authors believe all quadrats should be of the same shape and size to maintain the objectivity of the method in this step (Roig-Juñent, pers. comm.).

Distribution data: The geographic distribution of taxa that inhabit the study area is analyzed. The taxa may be of any taxonomic hierarchy and they must constitute a natural group, that is they must be monophyletic groups. But it is not necessary that the taxa should be related phylogenetically or ecologically.

Quadrats × taxa data matrix: A data matrix is constructed by assigning 1 if the taxon is present in the quadrat or 0 if it is absent (Fig. 7.1b). To root the cladogram a hypothetical quadrat is added in which it is considered that all the taxa are absent (0 in all the quadrats).

Matrix analysis using a maximum parsimony algorithm: The data matrix is analyzed by means of any software used for a cladistic analysis, for example, PAUP, Hennig86, or NONA. If more than an equally parsimonious tree is obtained, a strict consensus tree is constructed that shows the groups that are present in all the resulting trees (Fig. 7.1c).

Selection of areas of endemism: Only those groups of quadrats that form a monophyletic clade sustained by the presence of more than one taxon are considered. The selected quadrats are drawn on a map, defining the

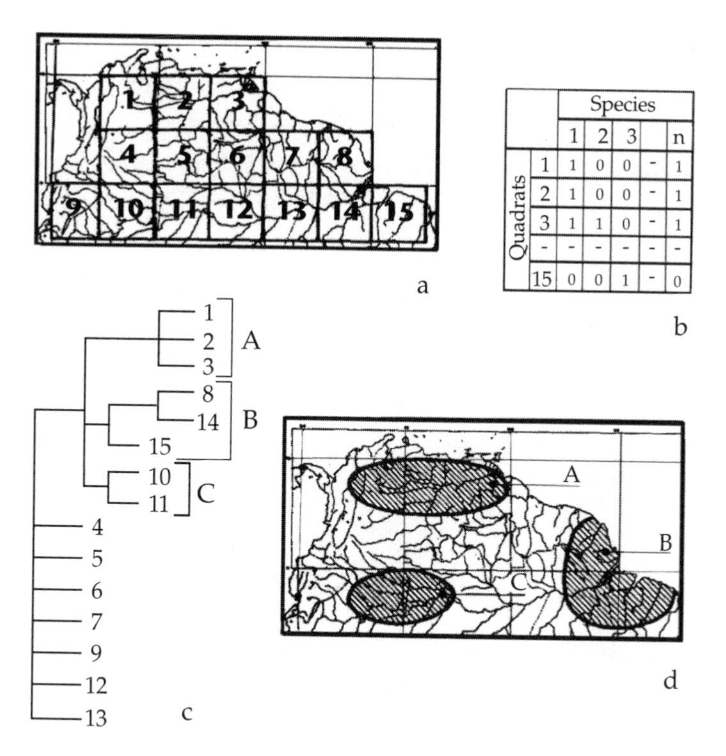

FIGURE 7.1. PAE based on quadrats. *(a)* Map of northern South America showing 15 quadrats; *(b)* data matrix (quadrats × species); *(c)* cladogram obtained applying a maximum parsimony algorithm; *(d)* areas of endemism (A, B, and C) delimited following the cladogram groups.

limits in function of the real distributions of those taxa that sustain each clade (Fig. 7.1d).

The patterns of nested areas obtained by means of PAE based on quadrats are useful from the point of view of biodiversity conservation (see chapter 13), because the minor units obtained include species with larger distributions and also those with restricted distribution (Posadas & Miranda-Esquivel, 1999).

The main criticism of PAE is that it ignores cladistic relationships among taxa, considering only their distributions (Humphries, 1989, 2000).

Some authors (for example, Craw, 1988a; Cracraft, 1991; Myers, 1991; Morrone, 1994b) incorporate cladistic information in the matrix, using supraspecific natural groups (containing two or more species). Yet, in PAE based on areas of endemism, adding cladistic information on supraspecific taxa is a misuse of the primary Brooks parsimony analysis (Crisci et al., 2000).

For the panbiogeographers (for example, Craw et al., 1999), the PAE based on localities and the PAE based on areas of endemism are related conceptually to panbiogeography. Moreover, Smith (1992) has even stated that PAE is a method that expresses panbiogeographic tracks in a hierarchic scheme; recently, his idea was empirically applied by Morrone and Márquez (2001).

CASE STUDY 1: PAE IN THE ANDEAN SUBREGION

The Andean subregion comprises the Andean highlands from Ecuador and Venezuela extending along the cordillera to southern Chile, following the Pacific coast, and including the Patagonian steppe from southern Argentina up to the Atlantic coast. This subregion's biota has complex relations with those of other temperate areas of South America, Australia, New Zealand, and other minor islands of the Austral hemisphere. To delimit areas of endemism within this subregion the parsimony analysis of endemicity (PAE) based on quadrats (Morrone, 1994a) was applied (Posadas et al., 1997).

Delimiting the area: The Andean subregion was considered in its totality (the subregion was described in the first case study in chapter 5).

Selecting the taxa: The distribution of 160 species of vascular plants of the subregion was analyzed. The distributional data were taken from revisions and monographs.

Elaborating the grid: The Andean subregion map was divided into 30 quadrats, whose size depended on the resolution level expected for each area, thus, the quadrats situated south of 34°S were smaller because this area is very biogeographically complex (Fig. 7.2).

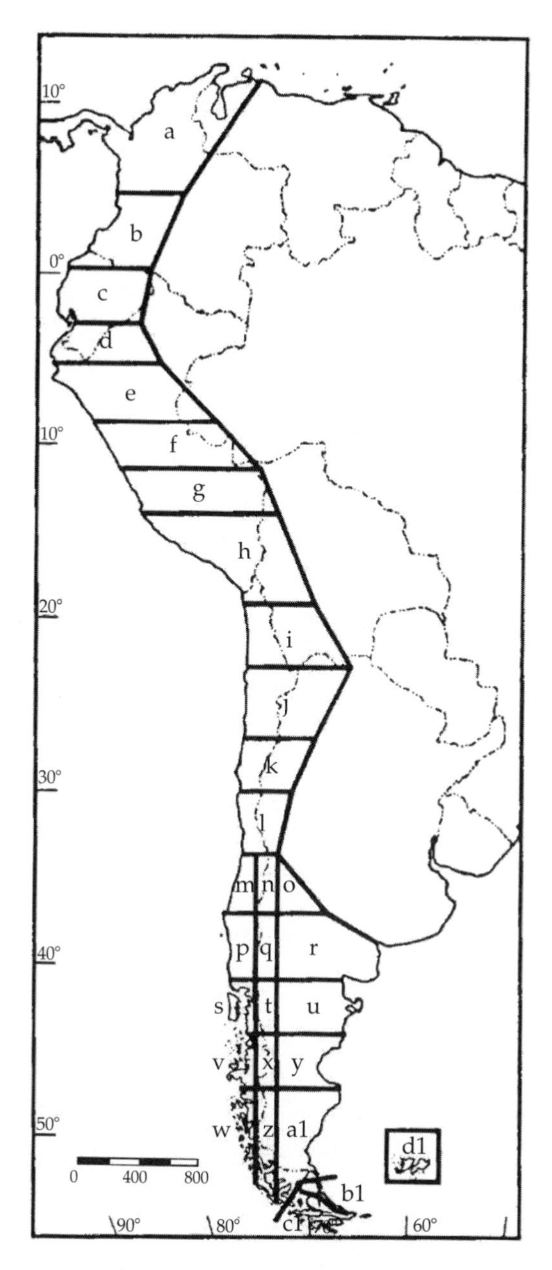

FIGURE 7.2. Partial map of South America showing the grid that divides the Andean Subregion into 30 quadrats. Each quadrat is indicated by a letter.

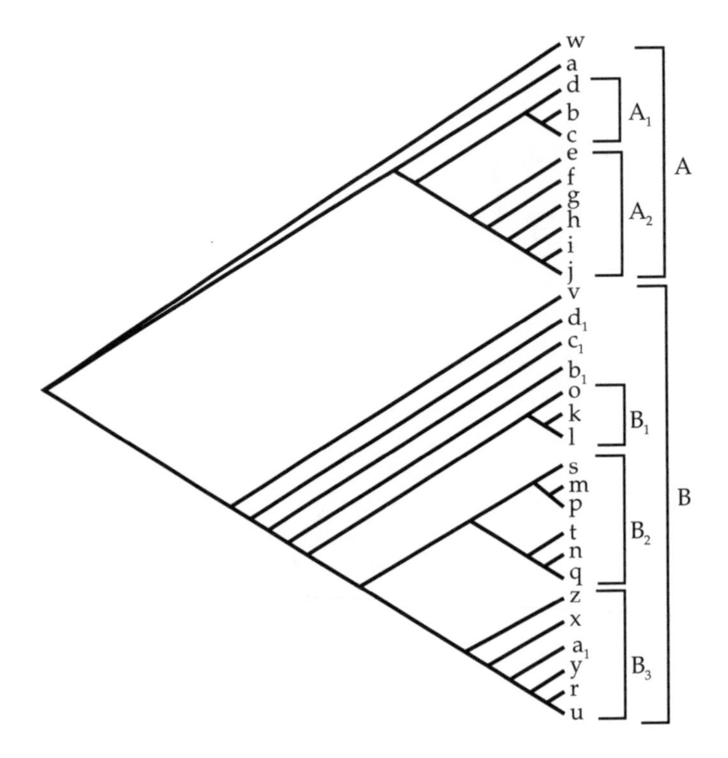

FIGURE 7.3. Cladogram obtained from the data matrix (quadrats × species); each terminal represents one quadrat. A and B represent major areas of endemism, and A_1, A_2, and B_1, B_2, and B_3 the minor areas of endemism included in A and B respectively.

Elaborating the matrix: A matrix of 30 quadrats × 160 species was constructed. The presence of a species in a quadrat was coded 1 and its absence 0. In the matrix was included a hypothetical quadrat coded 0 to root the cladogram.

Obtaining the cladograms: The matrix was analyzed with Hennig86, options mh*, bb*, and NONA programs, applying the options mult*25 and max*. Both analyses produced the same unique cladogram (Length= 366; CI= 0.41; RI= 0.61) (Fig. 7.3).

Delimiting areas of endemism: The cladograms showed two main areas, A and B. Within A two minor areas appeared (A_1 and A_2), whereas within B three minor areas appeared (B_1, B_2, and B_3) (Figs. 7.3 and 7.4).

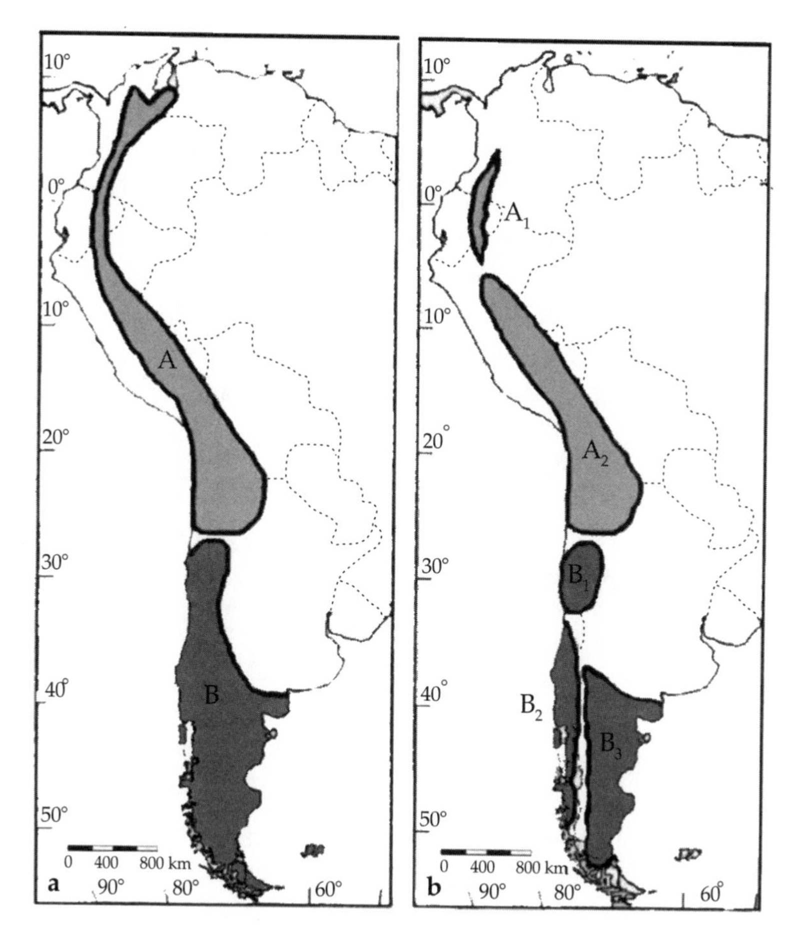

FIGURE 7.4. Areas of endemism in the Andean Subregion obtained applying PAE based on quadrats. *(a)* Areas A and B (shown in the cladogram of figure 7.3); *(b)* areas A_1, A_2, B_1, B_2, and B_3 (shown in the cladogram of figure 7.3). A_1 corresponds to the Páramo Province; A_2 corresponds to the Puna province; B_1 corresponds to the Central Chile Province sensu stricto; B_2 corresponds to the Subantarctic Province; and B_3 corresponds to the Patagonian Steppe Province.

The two main areas of endemism, A and B, represent the biogeographic provinces of the Páramo and the Puna (A), and the biogeographic provinces of Central Chile, the Subantarctic, and the Patagonian steppe (B). The minor areas in A are coincident with the Páramo (A_1) and the Puna (A_2) as they were defined by Cabrera and Willink (1973) and Morrone (1994b). These results prove that these biogeographic units constitute natural areas from a historical biogeographic point of view. Strictly speaking, the minor areas in B correspond to Central Chile (B_1; see the example below). The size of this area is smaller than it was in Cabrera and Willink's original description (1973). The area north of Central Chile remains related to the Puna zone (quadrats i and j), whereas the others are related to the Subantarctic province, represented by B_2. This complexity in the Central Chile province, with the Puna in its northern section and the Subantarctic in its most southern part, is coincident with the hypothesis of Morrone and colleagues (1997). B_3 represents the Patagonian steppe province (Fig. 7.4).

CASE STUDY 2: PAE IN CENTRAL CHILE

Most consider Central Chile to be located between 30°S and 37°S. The analysis of the distribution of its diverse animal and plant taxa shows a high percentage of endemic taxa and a close relationship between its austral zone and the Subantarctic province. An analysis of cladistic biogeography combined with PAE based on quadrats was carried out to redefine the Central Chilean province (Morrone et al., 1997).

Delimiting the area: The area known as Central Chile (Cabrera & Willink, 1973) was divided into minor areas on the basis of repeated patterns of distributions of the species of seven genera of Asteraceae (132 species), one genus of Buprestidae (Coleoptera, one species), three genera of Curculionidae (Coleoptera, 53 species), and two genera of Gnaphosidae (Araneae, 13 species). Four areas of endemism from Central Chile were recognized: Coquimbo (Co), Santiago (Sa), Curicó (Cu), and Ñuble (Ñu). The Subantarctic province (SUB) was used to root the tree (Fig. 7.5).

Obtaining taxon cladograms: The following cladograms were obtained for the analysis: *Triptilion* (Asteraceae, seven species), *Calopappus-Nassauvia* sect. *Panargyrum* (Asteraceae, six species), *Leucheria amoena* species group (Asteraceae, eight species), *Leucheria cerberoana* species group (Asteraceae, 12 species), *Mendizabalia* (Buprestidae, two species, one with two subspecies), *Listroderes nodifer* species group (Curculionidae, five species), *Listroderes curvipes* species group (Curculionidae, three species), *Puranius* (Curculionidae, 12 species), *Apodrassodes* (Gnaphosidae, three species), and *Echemoides chilensis* species group (Gnaphosidae, five species).

Obtaining area cladograms: In the aforementioned cladograms the taxa were replaced by the areas that they inhabit and the corresponding area cladograms were obtained.

Obtaining general area cladograms: Four different cladistic biogeographic methods were used: a) component analysis, assumptions 0, 1, and 2; b) primary Brooks parsimony analysis, using the program Hennig86, option ie*; c) three-area statements, using the programs TAS and Hennig86; d) paralogy-free subtrees, assumption 2, using the programs TASS and Hennig86. Through the application of these methods, five general area cladograms were obtained. From them the most parsimonious (minimum error items; see chapter 6) (Fig. 7.5) was selected by means of the fit option of the COMPONENT 1.5 program, the one that implies the minimum number of extinctions and dispersals.

PAE based on quadrats: The Central Chile area (30°–37°S) and the northern part of the Subantarctic province were divided into 13 units (A–M), each one degree of latitude (Fig. 7.5). A data matrix of 13 quadrats × 67 taxa was constructed. The presence of one species in one quadrat was coded 1 and its absence 0. In the matrix, a hypothetical quadrat all coded with 0 was included to root the tree. Hennig86, options ie* and successive weighting, was applied to the matrix, resulting in a sole cladogram.

The general area cladogram (Fig. 7.5) shows that a first vicariant event separated the northern areas (Co–Sa) from the southern ones (Cu, Ñu, and SUB); a second event separated Co from Sa; whereas within the southern areas Cu separated first and the last event separated Ñu from SUB. According to these hypotheses the four areas of Central Chile do not constitute a natural or monophyletic group because Cu and Ñu are cladistically more related to SUB than to the sister areas Co–Sa. The cladogram obtained with PAE based on quadrats (Fig. 7.5) is completely coincident with the general area cladogram. It shows the same relation-

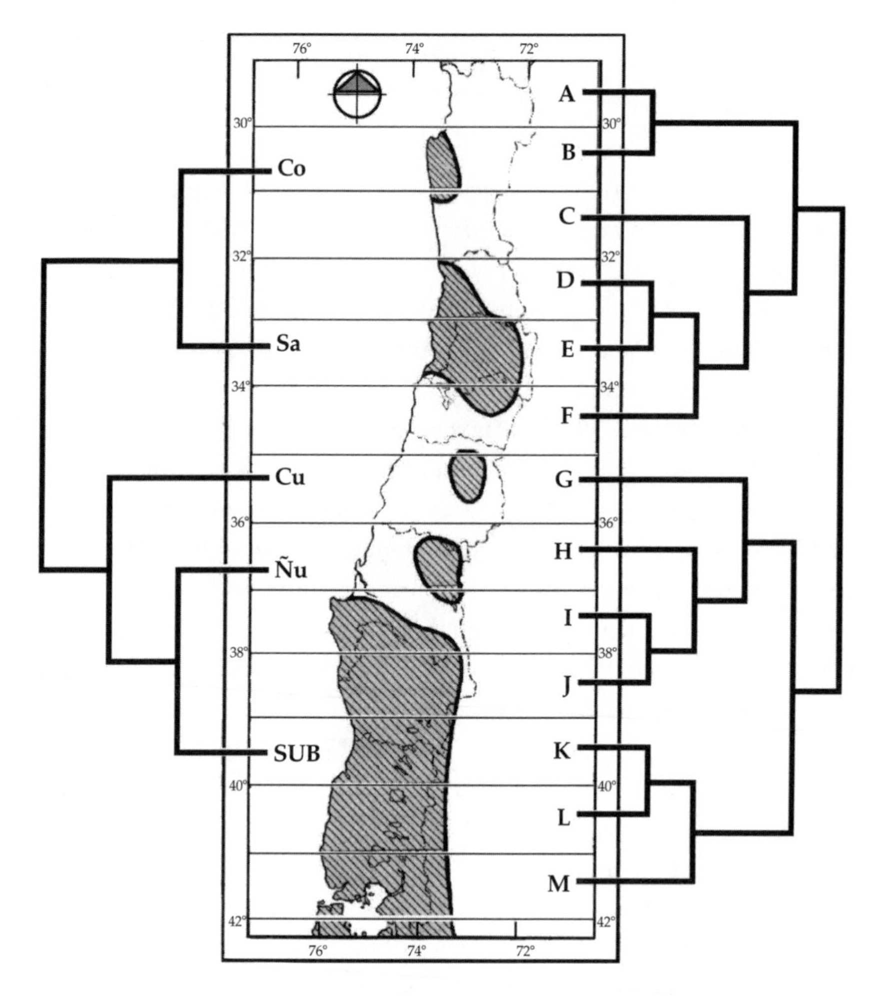

FIGURE 7.5. Map of Central Chile and northern Subantarctic showing the grid used to apply PAE based on quadrats (quadrats indicated by letters A to M). Oblique lines indicate the areas of endemism obtained. The general area cladogram obtained applying cladistic biogeographic techniques is shown on the left. The cladogram obtained applying PAE is shown on the right. Co = Coquimbo; Sa = Santiago; Cu = Curicó; Ñu = Ñuble; SUB = Subantarctic.

ships among the three geographic unit groups: one northern group (A–F), an intermediate group (G–J), and a southern group (K–M). The results suggest that the area traditionally known as Central Chile contains two sections, a northern section (Co–Sa, 30°–34°S, A–F) with a great number of endemic taxa, and a southern section (Cu–Ñu, south of 34°S, G–J) with endemic taxa and a great number of Subantarctic elements. The southern section is much more related to the Subantarctic province than to the northern section of Central Chile, as the previous case study also suggests.

RESEARCH USING PAE

Among the empirical applications that use PAE, we can mention Conran (1995) on the plant group Liliiflorae throughout the world; Morrone and Lopretto (1995) on southern South America, studying Decapoda; Morrone and Coscarón (1996) on America, studying Heteroptera; Posadas (1996) on the southernmost part of South America, studying vascular plants; Emerson and colleagues (1997) on New Zealand, studying Lepidoptera; Bates and colleagues (1998) on Neotropical lowlands, studying passerine birds; Linder and Mann (1998) on South Africa, studying Cape flora; Morrone (1998) on Udvardy's Insulantarctic province, studying weevils (Coleoptera); Morrone and colleagues (1999) on Mexico; Ron (2000) on Neotropical lowland rainforest, studying vertebrate groups; Marino and colleagues (2001) on southern South America, studying the insect family Ceratopogonidae (Diptera); Bisconti and colleagues (2001) on Galapagos terrestrial biota; Luna and Alcántara (2001) on Mexico; Trejo-Torres and Ackerman (2001) on Antilles, Florida, Yucatan, and the Guianas, studying orchids; and García-Barros and colleagues (2002) on animal and plant taxa from the Ibero-Balearic region.

8

EVENT-BASED METHODS

THE SO-CALLED event-based methods have lately been acknowledged in historical biogeography. Contrary to most other historical biogeographical methods, the event-based methods postulate explicit models of the processes that affect the geographic distribution of living beings. The different types of processes (dispersal, extinction, geographic paralogy, or duplication and vicariance) are identified and assigned values of benefit or cost under an explicit model of functioning in nature. Consequently, the history of a taxon's distribution is inferred on the basis of phylogenetic information and by applying the criterion of maximum benefits and minimum costs respecting all or only some of these processes. Those who propose these methods (for example, Page, 1994a) postulate a functioning model so as to have an optimality criterion (for example, minimum cost), an advantage over methods applying algorithms without defining optimality criteria.

These methods are also applied in coevolution studies of molecular systematics (organism-gene) and of parasitology (host–parasite). This is possible because there are basic similarities among the processes in these two associations and in the area-taxon relationship in historical biogeography. Page and Charleston (1998) postulate that in each of these three associations an "associate" is tied to the events that happen in the "host." Table 8.1 compares equivalent events in these three associations and their corresponding components. Even if the events of these associations are not homologous, they result in similar cladistic patterns. For instance, the duplication phenomenon of a gene within an organism generates a pattern similar to the speciation of a parasite in a host with no speciation of the latter, or the same pattern of sympatric speciation of a taxon in a determined area (Page & Charleston, 1998) (Fig. 8.1). In consequence, the molecular systematists, the parasitologists, and the historical biogeographers face a common problem, to reconstruct the history between an associate (gene, parasite, or taxon) and its host (organism, host, or area) (Page, 1994a). The three disciplines use the information contained in the cladograms of the hosts and associates to reconstruct these histories. Methodological developments in one discipline are then ultimately adopted by the others.

No fewer than six event-based methods have been proposed in recent years. In this chapter we will present five of them: reconciled trees or maximum cospeciation (MC), dispersal-vicariance analysis (DIVA), jungles, the Bayesian approach to cospeciation, and the combined method.

RECONCILED TREES OR MAXIMUM COSPECIATION

The concept of reconciled trees appears independently in molecular systematics, parasitology, and biogeography (Page 1994a, 2000; Patterson et al., 2000). Rod Page (1994a) proposed a method to reconstruct the host-associate history in biogeography. Consequently, we will use the nomenclature corresponding to this discipline.

Page (1994a) recognized four kinds of events. In the first, geographic paralogy, there is sympatric speciation. The associate (the taxon) suffers speciation with no change in the host (the area) (see Fig. 8.1). The second event is dispersal, in which the taxon colonizes a new area (Fig. 8.2). The

Table 8.1. Process equivalences among different historical associations, including different nomenclatures used by different authors.

	Association			Process		
	Host	Associate	Codivergence	Duplication	Horizontal transfer	Sorting event
Molecular systematics	Organism	Gene	Simultaneous speciation and gene duplication = interspecific coalescence = ortologous genes	Paralogy = gene duplication without speciation = deep coalescence = paralogous genes	Gene transfer	Gene loss = lineage sorting
Parasitology	Host	Parasite	Cospeciation = successive specialization = host tracking	Parasite speciation without host speciation	Host-switching = colonization	Parasite extinction = exclusion
Historical biogeography	Area	Taxon	Vicariance	Sympatry = taxon speciation without area vicariance = geographic paralogy = redundancy = multiple lineages	Dispersal	Extinction

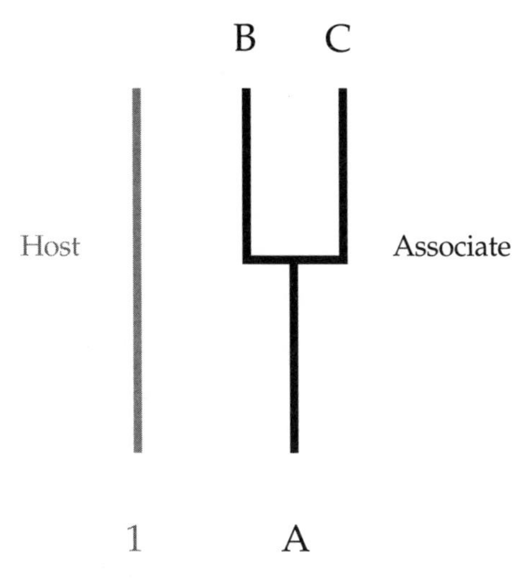

FIGURE 8.1. Cladistic pattern generated by a duplication event. In biogeography, duplication corresponds to a sympatric speciation event. The host corresponds to an area (1) and the associate is an organism (A) that has been speciated (B and C) without changes in the original distribution.

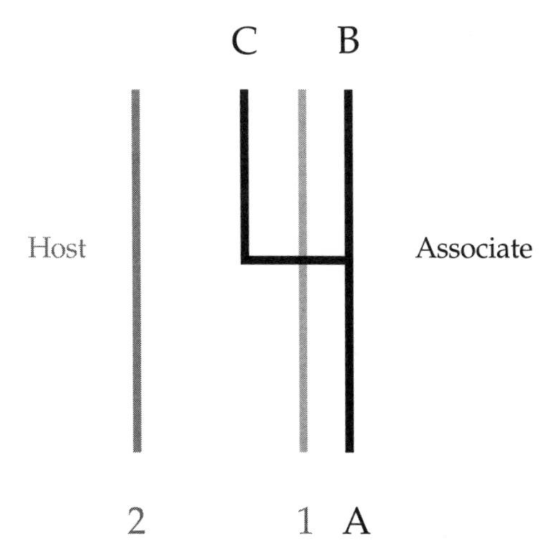

FIGURE 8.2. Cladistic pattern generated by a horizontal transfer event. In biogeography, horizontal transfer is equivalent to a dispersal event. Hosts (1 and 2) represent two areas not related each other; associates (organisms A, B, and C) have dispersed, then taxon C, which represents the sister group of taxon B, has dispersed to area 2, which is not related to the ancestral area of the taxon.

third is extinction, in which the taxon disappears from an area (Fig. 8.3). Sometimes a lack of sampling leads researchers to mistakenly believe there is an extinction. Finally, there is vicariance, in which there is allopatric speciation—the taxon undergoes speciation together with a divergence of the areas (Fig. 8.4).

MC maximizes the number of events of vicariance (codivergence), minimizes duplication and extinction, and forbids dispersal. Subsequently, Page (1994b) proposed a method to incorporate dispersal. To identify the taxa that have dispersed, each taxon must be deleted by turn and a reconciled tree for the remaining taxa must be generated. Those taxa whose deletion greatly increases the congruence between the area cladograms and the taxon cladograms have a high probability of dispersal.

This technique includes five steps:

1. Select the host and associate cladograms;
2. superpose each node of the associate cladogram on the host cladogram;
3. assume that cospeciation is maximum and do not consider dispersal;
4. ascribe the differences between the cladograms to the events of extinction or duplication, or to both events simultaneously; and
5. choose a reconstruction that implies a minor cost (i.e., that which contains the maximum cospeciation and minimizes the remaining events).

The algorithms to obtain reconciled trees are implemented in COMPONENT version 2.0 (Page, 1993).

Usually, the general area cladogram (host cladogram) is not known beforehand in biogeography. The way to proceed in this case is to reconcile the associate tree or trees (area cladograms) with each possible general area cladogram. If the number of considered areas is low, all the possible general area cladograms are enumerated. If the number of considered areas is high, an heuristic search takes place (Page, 1994a). In this

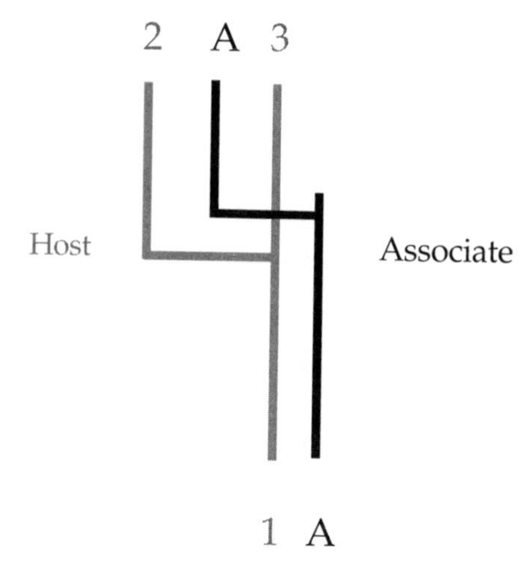

FIGURE 8.3. Cladistic pattern generated by a sorting event. In biogeography, a sorting event is equivalent to an extinction event. The sister group of taxon A (associate) has become extinct and, consequently, is not present in area 3 (host).

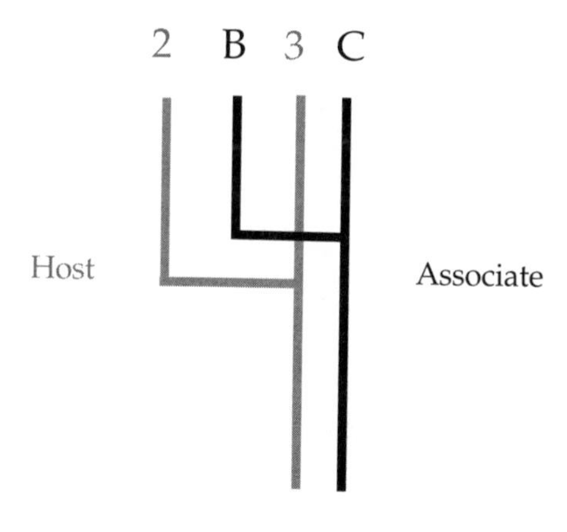

FIGURE 8.4. Cladistic pattern generated by a codivergence event. In biogeography, codivergence is equivalent to a vicariant event. The organisms B and C (associates) have suffered allopatric speciation as a response to a process of fragmentation of area 1 because of the appearance of a barrier. Such a process results in the subsequent differentiation of two areas, 2 and 3 (hosts). Associates b and c are relics of a larger clade of three associates, one of which (d′) on host D is now extinct. Associate d is a relic of a clade of three associates, two of which (b′, c′) are now extinct,

way, the host phylogeny that shows a maximum vicariance in respect to those of the associates is considered to be the general area cladogram.

Figure 8.5a shows an example of a reconciled tree between an area cladogram and its general area cladogram. This represents the simplest situation possible—the codivergence between both associates is at the maximum. Figure 8.5b shows a more complex example of a similar situation where a duplication is needed (node f′ in the reconciled cladogram) to reconcile both trees.

This method applies the assumption of one host per parasite that in biogeography corresponds to the assumption of one area per species. It means that the ancestral species are allowed to occur in single areas or in multiple areas, from which it has been postulated that they form a contiguous region in the past, according to the general area cladogram.

MC is a useful method to analyze whether in an association there exists more vicariance than the one that would exist at random. The number of vicariant events obtained according to the original data are determined, and then the taxa associated with the areas are exchanged at random, or the area cladograms or the general area cladograms are exchanged at random. The maximum number of vicariant events for the randomized data is calculated, and the proceeding is repeated several times. If the value observed for the real data is higher than the value calculated for the randomized exchange in 95 percent of the replications, the hypothesis of an association at random between the taxon phylogenies and that of the areas can be rejected (Ronquist, 1998).

The MC optimization algorithms also work with alternative mapping of host nodes onto parasite nodes, in an attempt to find the best fit between the trees.

Fredrik Ronquist (1998) presents a formalization of this method from the point of view of three-dimensional matrices of benefit. In this formalization a value of -1 (benefit value) is assigned to each vicariance event; the duplications and extinctions have no cost nor benefit (value 0). The model does not allow dispersals so a value of infinite (prohibition value)

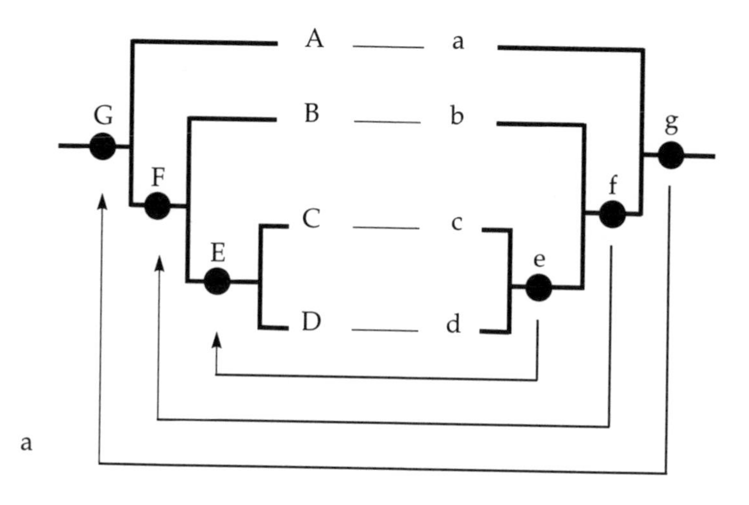

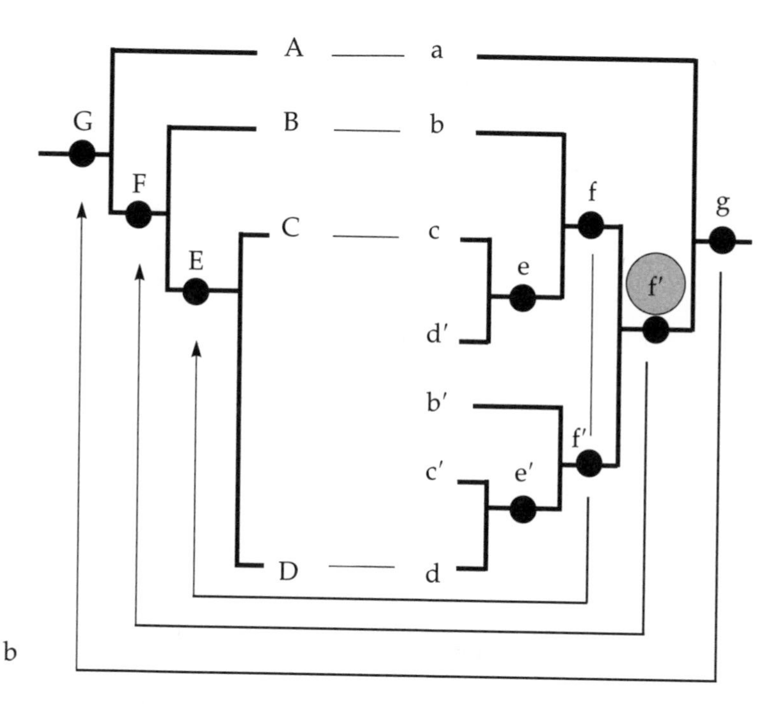

FIGURE 8.5. Reconciled trees or maximum cospeciation method. *(a)* A trivial example in which the cospeciation between the host and the associate trees is complete. *(b)* In the second example it is necessary to postulate a duplication event, represented by f′ node, to reconcile the host and its associate tree. Associates b and c are relics of a larger clade of three associates, one of which (d′) on host D is now extinct. Associate d is a relic of a clade of three associates, two of which (b′, c′) are now extinct.

is assigned to them. Synthesizing, the model maximizes the cospeciation events (vicariance), minimizes the duplications and extinctions, and does not consider the dispersals.

DISPERSAL-VICARIANCE ANALYSIS (DIVA)

This method was proposed by Ronquist (1997b), based on ideas developed by Ronquist and Nylin (1990), to study associations between organisms (host-parasite). Such associations were called a "coevolutionary two-dimensional cost matrix." DIVA reconstructs the taxon ancestral distribution on the basis of a simple biogeographic model. It is possible to apply this method to the biogeographic study of a taxon even when no general area cladogram exists. Furthermore, DIVA allows the reconstruction of biogeographic scenarios that include the possibility of reticulate relations existing among the areas and not solely hierarchical relations, as with all cladistic biogeographical methods. Dispersal-vicariance analysis works by assuming that the distributions of the species and their ancestors may be described in terms of a set of area units. The dispersal-vicariance method reconstructs the ancestral distribution ranges at each of the internal nodes of the cladogram by means of a set of optimization rules and costs for dispersal and extinction events (adding or losing areas). The justification for this approach is that a pure vicariance analysis assumes that the ancestral range of the clade is the total range that its modern species occupy (Bremer, 1992), which may be considered unlikely. There are two optimization rules: The ancestral node may not contain an area that is unoccupied by any descendant; and the ancestral node must contain at least one area from the distributions of each of the two descendant nodes.

Where these rules leave ambiguous optimizations, a cost is assigned to extinction and dispersal events (which add or lose areas), whereas allopatric and sympatric speciation carry no cost (redundant distributions and vicariance) (Linder, 1999). These costs are calculated in three steps, with down-pass, up-pass, and final-pass operations implemented in the DIVA program (Ronquist, 1996; see below).

The cost matrix, constructed from the distributional and phylogenetic data, is three-dimensional. The premises for the construction of these matrices are the following:

Vicariance events have a null cost (= 0). Speciation is assumed to be by vicariance, separating a wide distribution into two mutually exclusive sets of areas.

Duplication events receive a cost of 0. Species occurring in a single area may speciate within the area by allopatric (or possibly sympatric) speciation, giving rise to two descendants occurring in the same area.

Dispersal events receive a cost of 1 per each area unit added to distribution.

Extinction events receive a cost of 1 per each area unit deleted from a distribution.

Thus, the distributions are explained in a way that implies the least possible cost.

Ronquist (1996) developed DIVA version 1.1 for the application of this method. As a result, it is now possible to reconstruct the ancestral distributions for each node. Furthermore, a series of statistics that show the frequencies of possible events among two or more areas, may be obtained, separating dispersal and vicariance. These frequencies could be transformed in percentage values considering the total frequency of events (dispersal or vicariance) as 100 percent, then calculating the proportion in which each particular event contributes to this total (Posadas & Morrone, 2001).

In a later paper, Ronquist (1997a) proposed a modification of his method, which he named "constrained DIVA." In this new approach he distinguishes between random dispersals (those that imply that a taxon passes through a barrier) and predictable dispersals (when the dispersal event is given because a barrier disappears).

The cost matrix in this case is constructed according to the following rules:

The vicariance events receive a benefit value, equal to −1 cost;
the duplication events receive a 0 cost;
the extinction events receive a cost of 1;
the random dispersal events receive a cost of 1; and
the predictable dispersal events receive a benefit value, equal to −1 cost.

Ronquist (1998) emphasizes that the models proposed thus far are oversimplified. This notwithstanding, he predicts that the three-dimensional cost matrix framework is powerful enough to allow optimization based on virtually any conceivable coevolutionary or biogeographic model, regardless of its complexity. For instance, in biogeographic inferences it is possible to assign cost to dispersal events relative to the distance traversed (Ronquist, 1997a,b), or impose constraints based on time-segmented reticulate biogeographic scenarios.

Table 8.2 compares the values assigned to the different events in

Table 8.2. Cost values assigned to each event in those methods that use three-dimensional cost-benefit matrices. Values assigned to maximum cospeciation (MC) correspond to Ronquist's (1998) formalization of this method.

	Cost values		
Event	MC	DIVA	Constrained DIVA
Vicariance	−1	0	−1
Duplication	0	0	0
Extinction	0	1	1
Random dispersal	∞	1	1
Predictable dispersal	Not considered	Not considered	−1

the methods that use three-dimensional matrices (MC, DIVA, and constrained DIVA).

JUNGLES

As seen before, the reconciled trees method (MC) has some limitations. For instance, it does not consider host switching (Page & Charleston, 1998). Host switching is a transfer of an associate lineage from a source host toward another destination host that is not the immediate descendant of the source host. The horizontal transfer of genes, host switching, and biogeographic dispersal are examples of this. To postulate a host switch it is necessary that the source and destination hosts are contemporary, then the relative ages of different host lineages must be considered. Not bearing in mind this temporal limitation results in a postulate of transferences that are mutually incompatible. Charleston (1998) developed a solution to this problem that uses a mathematical structure called "jungle" that contains all the possible partial orderings in which the associate tree may be tracked in the host tree, considering the codivergence, duplication, sorting, and host-switching processes, and all the existing known associations. Once the costs are calculated for each of these processes, it is possible to find the subgraph or subgraphs that correspond to the least costly reconstructions of the association history (that is, the most parsimonious one). To our knowledge, applications of this method in biogeography do not exist.

BAYESIAN APPROACH TO COSPECIATION

Similar to other event-based methods, the Bayesian approach to cospeciation (Huelsenbeck et al., 2000a) was developed to be applied specifically to coevolutive studies of host-parasite associations. The objective of the method is to infer the frequency of host switching and the lineages involved in host switches using a statistical method that assumes a stochastic model.

According to Huelsenbeck and colleagues (2000a), in the traditional works the problem of inferring where host switching events occurred relied on maximizing the number of cospeciation events relative to the host switching and sorting events that are needed to explain the differences between the host and the parasite trees (Page, 1994a). This type of analysis assumes that the host and parasite phylogenies are estimated without error, and that the host switching rate is low enough that all switching events are evident from an analysis of particular regions of disagreement between host and parasite phylogenies. A Bayesian estimation can be used in models where host switching events are assumed to occur at a constant rate over the entire evolutionary history of associated hosts and parasites. The method provides information on the probability that an event of host switching is associated with a particular pair of branches, and reduces the probability that a particular phylogeny may be overturned if a reexamination of the group results in a different tree. This is because each considers all potential trees, weighted according to the probability that each is correct in testing an evolutionary hypothesis. This method does not take into account such biological events as duplication and host extinction.

As we suggested at the beginning of this chapter, there are basic similarities among the different processes given in host-parasite association and in area-taxon association. Huelsenbeck and colleagues (2000a) postulated that biogeography might usefully be studied in a framework similar to the one adopted in the Bayesian approach, in which the area cladogram is fixed and the multiple species trees on the area cladogram are variable.

COMBINED METHOD

In recent research (Miranda-Esquivel, 1999; Posadas & Morrone, 2001) a utilization of combined approaches has been proposed. This synthesis consists of evaluating general area cladograms obtained by techniques in cladistic biogeography (for example, BPA) or event-based methods (for example, reconciled trees) in light of the statistics given by DIVA. The utility of combining these approaches results from the fact that construction techniques of general area cladograms tend to optimize only the events due to vicariance, because those events result in hierarchical patterns that can be represented by cladograms. Thus, the effects of the dispersal that generates a pattern of reticulate relations, impossible to show in tree diagrams, remain hidden. Furthermore, the dispersal events can be expressed in taxon distributional patterns, which generate biogeographic noise in the cladograms, generating relations between areas that do not share a common history. In those methods that only give hierarchical relations between the areas, groups of areas may arise whose relations are actually a result of dispersal rather than a common history.

By analyzing the reconstruction frequencies for biogeographic scenarios obtained from DIVA, it is possible to evaluate if relations in the general area cladograms are under the influence of dispersal or if they are really vicariance effects.

THE PROBLEM OF METRICITY

An important point that concerns three-dimensional cost matrices (for example, DIVA) and the reconciled trees method is the violation of metricity. A numerical function $d(xy)$ of pairs of points from a set E is metric if it satisfies the following conditions:

$d(xy) = d(yx) \geq 0$ (symmetry);

$d(xz) \leq d(xy) + d(yz)$ (triangular inequality);

If $d(xy) = 0$ then $x = y$ (distinguishability of nonidentical elements);

$d(xx) = 0$ (undistinguishability of identical elements).

This means that metricity is violated if 0 cost is not assigned to duplications, or if some of the events have benefit values (negative values) instead of cost values. Thus, the three-dimensional cost matrices required by DIVA and MC (in the version formalized by Ronquist, 1998) are nonmetric. Violation of metricity is an inevitable consequence if vicariance is considered as an event, which is more probable than duplication. The duplications imply a parasite speciation independent of the host in coevolution and a sympatric speciation in historical biogeography. This apparently makes duplications less probable than vicariance, due to which metricity imposes limitations that many biologists do not accept in the coevolutive or biogeographic inferences (Ronquist, 1998). Notwithstanding, mathematicians consider that violation of metricity implies the existence of an abstract nonmetric space in the calculus whose geometric properties are very difficult to explore and is therefore more convenient to avoid (Williams & Dale, 1965).

CASE STUDY: COMBINED METHOD IN THE SUBANTARCTIC PROVINCE

A historical biogeographic analysis was conducted to establish the relations among the different districts that form the Subantarctic province and to establish its relation with the Central Chilean Province on the basis of the distributional patterns and phylogenetic information of several arthropod groups of the area (Posadas & Morrone, 2001). The data was analyzed following the combined method. The methods of primary Brooks parsimony analysis (BPA) and reconciled trees (MC) were applied to obtain general area cladograms that were then evaluated against the results given by DIVA.

Defining areas: The areas considered in the analysis were the Central Chilean province and the districts of the Subantarctic province: Maule, Valdivia, Magellanic Forest, Magellanic Moorland, and Malvinas Islands (Cabrera & Willink, 1973; Morrone, 1994b, 1996a; Morrone et al., 1997).

Obtaining taxonomic cladograms: Cladograms were obtained by the usual techniques of cladistics of the following genera: *Echemoides*

(Araneae: Gnaphosidae), *Aeghorhinus, Puranius, Germainiellus, Antarctobius, Falklandius* generic group, and *Rhyephenes* (Coleoptera: Curculionidae).

Resolving polytomies and obtaining area cladograms: As the programs COMPONENT 2.0 (Page, 1993) and DIVA 1.1 (Ronquist, 1996) only work with totally dichotomic trees, all the possibilities of resolutions for each polytomy were considered in the analysis. Afterward, in every cladogram the taxa were replaced by the areas in which they occur.

Obtaining general area cladograms: The data were analyzed using independently the primary BPA and MC techniques. To apply primary BPA (see chapter 6), an area × components matrix was constructed from all the mentioned cladograms. The resulting matrix was 6 areas × 100 components, which was analyzed applying a maximum parsimony algorithm by means of the program Hennig86. To apply MC, a data matrix corresponding to the area cladograms of all the mentioned taxa was constructed. The data were analyzed in the program COMPONENT 2.0 applying a heuristic search. Three searches were made: In the first only the losses were minimized, in the second the duplications were minimized, and in the third both criteria were minimized at the same time. During all the searches codivergence was maximized. Both analyses (BPA, MC) yielded the same general area cladogram (Fig. 8.6). It shows Central Chile as the sister group of the pair formed by the northern districts of the Subantarctic provinces (Maule and Valdivia), and these three groups together as the sister group of the three southern Subantarctic districts (Malvinas Islands, Magellanic Forest, and Magellanic Moorland).

Application of DIVA: The data matrix containing the phylogenetic and distributional information of all mentioned taxa was run on program DIVA 1.1. The option of evaluation of ambiguous events was used (that is, that in case of more than one possible reconstruction for each node, all the reconstructions were considered). All the cladograms were analyzed as a group.

The general area cladogram obtained by primary BPA and MC shows

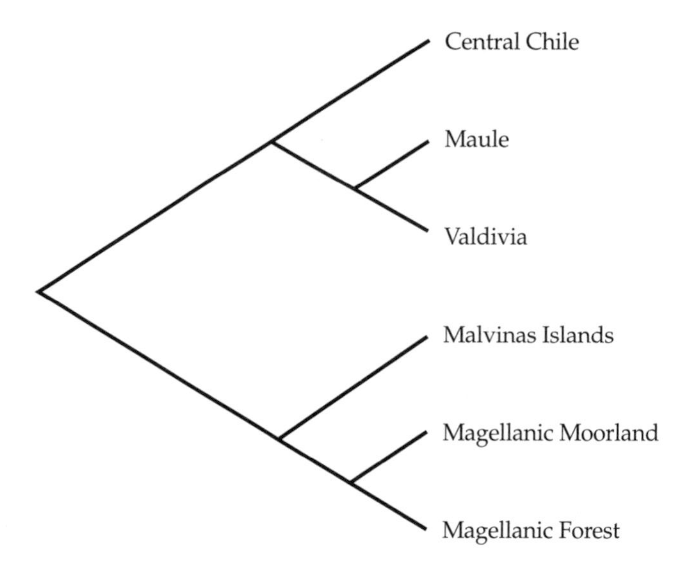

FIGURE 8.6. General area cladogram obtained applying primary BPA and reconciled trees methods. According to DIVA results, the closest relation between Central Chile and the northern Subantarctic provinces (Maule and Valdivia) could be due to dispersal events and not to vicariance events.

the biotas of the northern districts of the Subantarctic provinces and Central Chile to be closely related. Yet in view of the results of DIVA it is evident that this relation does not arise from vicariance events (common history) but from dispersal events, because 61 percent of the total dispersal events seem to have occurred in these three areas. Furthermore, the most frequent dispersal events always imply the Maule district, be it respecting Central Chile or Valdivia. Thus, the apparent complexity of this region's biota could be considered a result of the capacity of the biota to colonize new areas. In contrast, the relations exhibited by the Subantarctic southern districts can be contributed to vicariance events because DIVA shows that the dispersal frequencies in these areas are low, and the most frequent vicariance event is the separation stated in the general area cladogram of the Malvinas Islands from the pair formed by the Magellanic Forest and Magellanic Moorland.

RESEARCH USING EVENT-BASED METHODS

There are some empirical applications that use these techniques in historical biogeography. Among them are Nordlander and colleagues (1996) on cynipoid wasps (Hymenoptera); Beyra and Lavin (1999) on Leguminosae genus *Pictetia;* Fritsch (1999) on the widely distributed plant genus *Styrax;* Linder (1999) on the Danthonioid grass *Rytidosperma* from Australia; Miranda-Esquivel (1999) on the Simuliini tribe (Diptera) from America; Voelker (1999a) on the cosmopolitan passerine genus *Anthus;* Zink and colleagues (2000) on arid land birds from North America; Sanmartín and colleagues (2001) based on nonmarine animals from the Holarctic region; and Vinnersten and Bremer (2001) on the cosmopolitan plant order Liliales.

9

PHYLOGEOGRAPHY

THE INTRODUCTION of mtDNA data to population genetics in the late 1970s prompted a revolutionary shift in attitude toward historical, phylogenetic perspectives on intraspecific population structure. Because of the maternal, nonrecombining mode of mtDNA inheritance and rapid evolution in mtDNA sequence, the molecule often provides multiple alleles or haplotypes that can be ordered phylogenetically within a species, yielding intraspecific phylogenies (gene genealogies) interpretable as matriarchal components of the organismal pedigree (Avise, 1994). Due to the maternal, nonrecombining inheritance of mtDNA (at least in most species), all parts of the molecule share the same pattern of common ancestry. Furthermore, mtDNA clones and clades within many species have proved to be geographically localized. Such observations prompted introduction of the word "phylogeography" (Avise et al., 1987), which refers to the study of the principles and processes governing the geographic distributions of genealogical lineages, including those at the intraspecific level, on the basis of molecular data. Recently, Avise (2000) published a book that provides an exposition of the history, method, applications of, and prospects for phylogeography.

The use of mtDNA gene genealogies as well as the geographic information of the sampled populations allow the genetic structure of the populations to be evaluated (Hillis et al., 1996a). The amount and distribution of variation within and among populations depends on population sizes and rates of gene flow, both historical and contemporary. The most common pattern, at least in terrestrial vertebrates and higher plants, is to have less variation within than among geographic populations, indicating geographic population structuring.

Once the genetic structure of populations is analyzed it is possible to seek history's influence on such genetic structure. The tendency for greater variation among populations than within populations makes mtDNA usable for estimating phylogenies of populations. This in turn invites the investigation of patterns in historical biogeography, processes of long distance dispersal, past fragmentation, range expansion, and colonization. According to Ronquist (1997a), phylogeography is an approach to historical biogeography on an ecological scale of time.

One interesting approach to phylogeography is to compare all codistributed species (sympatric) to see whether or not they exhibit congruent patterns. Incongruent patterns are strong evidence of diverging stories resulting from species' differences in response to barriers or selective gradients, levels of gene flow, rates of molecular evolution, effective population size, or generation time. That is, species may have been historically codistributed, but variable responses to historical events produced conflicting phylogeographic patterns. Species that have recently colonized an area might not exhibit phylogeographic patterns simply because of insufficient time in situ for differentiation and because vicariant events evident in some species simply preceded their arrival to the community. Comparison of individual phylogeographies offers insight into the recent histories of communities. If many species in a community exhibited evidence of common responses to historical events, one could infer a long history of co-association of the component species, and might then ask questions about coevolution (Zink, 1996).

One of the first phylogeographic applications of mtDNA data was a study of the pocket gopher (*Geomys pinetis*), which inhabits the southeastern United States. Analysis of 87 individuals from across the species distributional range by six restriction enzymes revealed 23 different mtDNA haplotypes. Most mtDNA haplotypes in these gophers are localized geographically, appearing only at one or a few adjacent collection sites. Furthermore, genetically related clones tended to be geographically contiguous or overlapping, and a major gap in the matriarchal phylogeny exhibited a strong geographic orientation distinguishing eastern from western populations.

Population subdivision characterized by localized genealogical structure and significant mtDNA phylogenetic gaps across a species range subsequently have been reported in a wide variety of animal species (for example, mammals ranging from mice to whales, birds, reptiles, amphibians, freshwater and marine fishes, insects, snails, and crabs). In general, differences in organismal mobility and in environmental fragmentation appear to exert important influences on patterns of mtDNA phylogeographic structure.

Once the major evolutionary patterns of animal mtDNA were revealed, it might be supposed that they would apply to plant mtDNA and to other cytoplasmic genomes as well. But such has not proved to be the case. Plant mtDNA evolves quickly with respect to gene order but slowly in nucleotide sequence. This fact plus technical difficulties of assay have conspired to limit the applications of plant mtDNA for intraspecific phylogeography. Nonetheless, chloroplast DNA (cpDNA) variation is present and known to be structured geographically in several plant species (Avise, 2000). Chloroplast DNA is transmitted maternally in most plants, and the rate of evolution generally appears slow both in terms of primary nucleotide sequence and in terms of gene rearrangement (Avise, 1994). The chloroplast genome is far larger and more complex than the animal mitochondrial genome (Hillis et al., 1996a). However, some studies have uncovered considerable intraspecific cpDNA variation as well. The high levels of intraspecific variation, restricted vagility of seed propagules, and the presence of palynological databases make many plant species ideal model systems for phylogeographic studies. Analyses of several herbaceous species in southeastern North America provide examples of past fragmentation of biotas, range expansion, long distance dispersal, and colonization (Cruzan, 1999).

PHYLOGEOGRAPHIC HYPOTHESES AND COROLLARIES

In 1987, a comparative summary of molecular phylogeographic patterns suggested that historical biogeographic factors as well as contemporary ecologies and behaviors of organisms had played important roles in shaping the genetic architectures of extant species (Avise et al., 1987). These preliminary findings led to several intraspecific phylogeographic hypotheses and theoretical corollaries deemed worthy of further evaluation as new molecular and other evidence became available.

The three phylogeographic hypotheses for mitochondrial gene trees as formulated originally by Avise and colleagues (1987) are:

1. Most species are composed of geographic populations whose members occupy recognizable matrilineal branches of an extended intraspecific pedigree. Populations of most species display significant phylogeographic structure supported by mtDNA data.
2. Species with limited or "shallow" phylogeographic population structure have life histories conducive to dispersal and have occupied ranges free of firm, long-standing impediments to gene flow. Nonsubdivided, high-dispersal species may have limited phylogeographic structure.
3. Intraspecific monophyletic groups distinguished by large genealogical gaps usually arise from long-term extrinsic (biogeographic) barriers to gene flow. Major phylogeographic units within a species reflect long-term historical barriers to gene flow.

This last hypothesis includes four corollaries (discussed below) representing four aspects of genealogical concordance (Fig. 9.1).

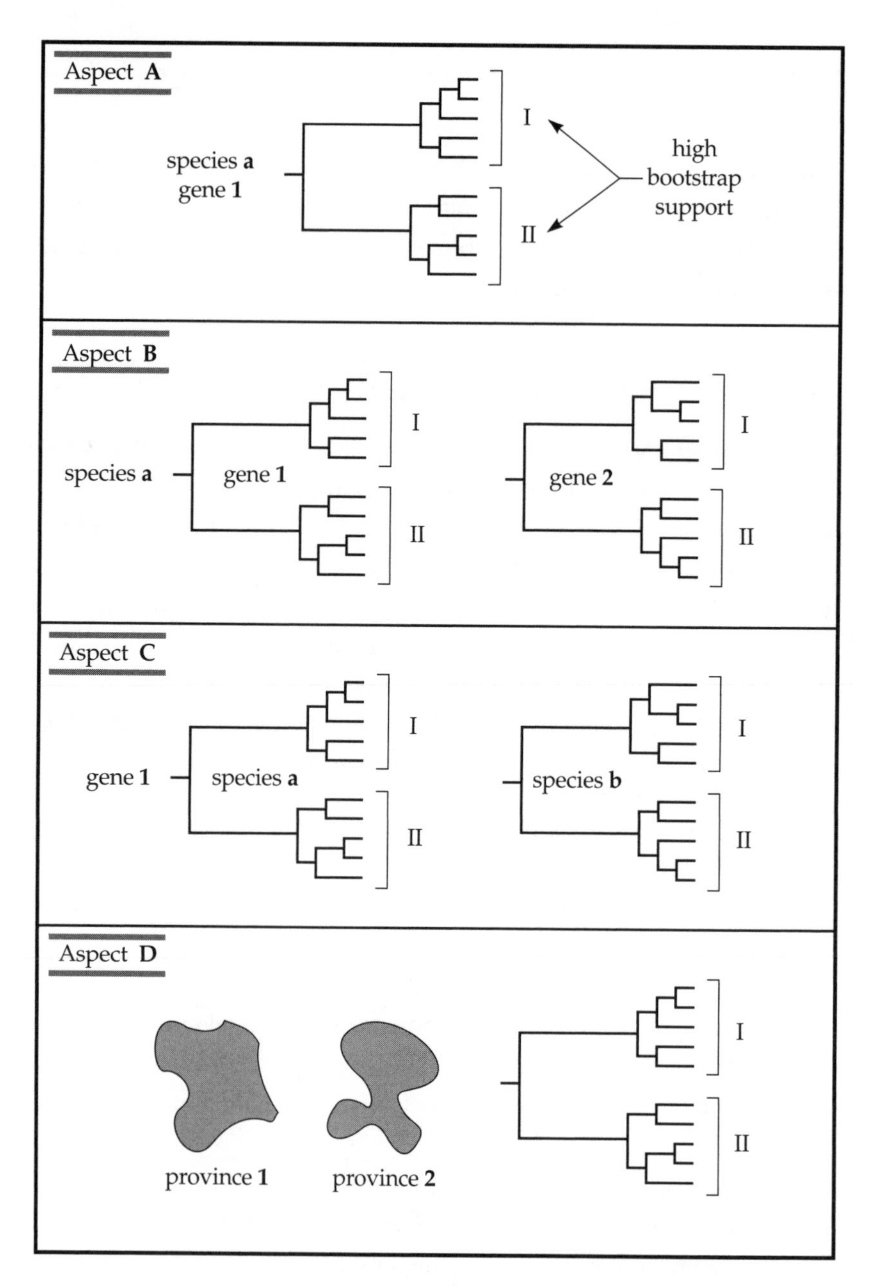

FIGURE 9.1. Phylogeography. Schematic presentation of four distinct aspects of genealogical concordance. I and II are distinctive phylogroups in a gene tree (modified from Avise, 2000).

Concordance across sequence characters within a gene (Agreement across characters within a gene): Every deep phylogenetic split in the intraspecific gene tree is supported concordantly by multiple diagnostic characters (nucleotides or restriction sites) within the mitochondrial genome. If this is not the case, such matrilineal separations would not be evident in the data analysis, nor would they receive significant phylogenetic support by criteria such as bootstrapping (Felsenstein, 1985).

Concordance in significant genealogical partitions across multiple genes within a species (Agreement across genes): Empirical examples show general agreement between deep phylogeographic topologies in multiple gene trees (such as mitochondrial and nuclear) within the species of interest. These deep branch separations characterize the same sets of geographic populations.

Concordance in the geography of gene-tree partitions across multiple codistributed species (Agreement across codistributed species): Several codistributed species with comparable natural histories or habitat requirements proved to be phylogeographically structured in similar fashion. In particular, divergent branches in the intraspecific gene trees might map consistently to the same geographic regions.

Concordance of gene-tree partitions with spatial boundaries between traditionally recognized biogeographic provinces (Agreement with other biogeographic data): An emerging generality from molecular phylogeography studies is that deeply separated phylogroups at the intraspecific level often are confined to biogeographic provinces or subprovinces as identified from traditional faunal lists.

NESTED CLADE ANALYSIS

Recently, statistical approaches that incorporate demographic-phylogenetic methods into phylogeography have been developed. Among them is the nested clade analysis (NCA) (Templeton, 2001), also called "cladistic nested analysis." This method involves an overlay of geography on an estimated gene tree in a rigorous statistical framework designed to measure the strength of any geography/phylogeny associations and to interpret the evolutionary processes responsible (Avise, 2000).

Population structure can be separated from population history when it is assessed through rigorous and objective statistical tests. This procedure provides greater statistical power and precision than traditional F-statistics, which do not use temporal information on allelic variation for detecting genetic and geographical associations. Two main statistics are estimated: the clade distance, which measures the geographical spread of a clade, and the nested clade distance, which measures how a clade is geographically distributed relative to other clades in the same higher-level nesting category.

NCA phylogeographic analysis and its inference criteria are discussed at length and a detailed example is given in Templeton and colleagues (1995). A validation of the inference criteria is given in Templeton (1998), and software that implements NCA (GeoDis; Posada et al., 2000) is available on the World Wide Web (see Appendix B).

CASE STUDY: PHYLOGEOGRAPHY OF RODENTS IN AMAZONIA

Phylogenetic relationships of populations of five genera of arboreal echimyd rodents (Caviomorpha) of the Amazon Basin were examined to reconstruct the biogeographic history of this region based on shared patterns for a number of small mammal taxa (da Silva & Patton, 1993). The extent to which populations of a series of independent lineages (species within genera) show concordant geographic patterns and time since divergence were assessed. Such concordance would imply a common underlying history.

Selection of taxa: Genera of the family Echimydae were chosen because they represent the most taxonomically rich and ecologically diverse group of living caviomorph rodents. Moreover, they constitute both the most abundant terrestrial and most diversified arboreal small mammals in the tropical forest community. Sixty-five individuals in five genera and nine species were examined: *Mesomys hispidus, M. stimulax, M.* sp., *Isothrix bistriata, I. pagurus, Makalata didelphoides, Dactylomis boliviensis, D. dactylinus,* and *Echimys chrysurus.*

Selection of area: Twenty-four localities within the Amazon Basin were sampled, in Venezuela (Amazonas), Perú (Amazonas), Brazil (Acre, Amazonas, Pará), and Bolivia (La Paz, Pando).

Molecular analysis: Mitochondrial sequences containing the cythochrome b gene (cyt b) were isolated via the polymerase chain reaction (PCR). Several software packages were used to analyze the sequence data. Pairwise comparisons to get percentage divergence values, the number of transition and transversion differences by codon position, and the proportions of each base at each codon position were performed. Assessments of phylogenetic relationships among observed cyt b haplotypes for each genus were derived from both parsimony and maximum likelihood distance analyses. A parsimony analysis was independently performed for each of the five genera represented in this study using PAUP. Distance-based analyses were done using PHYLIP; the DNADIST subroutine was used to calculate maximum likelihood distances.

Intraspecific nucleotide sequence divergence between mtDNA haplotypes of the cyt b gene is extensive (ranging up to 20 percent or more), and the mtDNA clades identified are strongly partitioned geographically. Both the degree of differentiation and the geographic patterning of the variation suggest that more than one species composes the Amazonian distribution of the currently recognized *Mesomys hispidus, Isothrix bistriata, Makalata didelphoides,* and *Dactylomys dactylinus.* The depth of divergence among several of the clades identified for these taxa suggests that sequence differentiation will be an extremely useful approach to the definition of species units in groups that have been historically very difficult to differentiate morphologically. There is general concordance in the geographic range of haplotype clades for each of these taxa, and the overall level of differentiation within them is largely equivalent. These observations suggest that a common vicariant history underlies the respective diversification of each genus. However, estimated times of divergence based on the rate of third position transversion substitutions for the major clades within each genus typically range above 1 million years. Thus, allopatric isolation precipitating divergence must have been considerably earlier than the late Pleistocene forest fragmentation events commonly invoked for the Amazonian biota.

RESEARCH USING PHYLOGEOGRAPHY

Currently, phylogeography is an extremely active discipline and this fact is reflected in the great number of recent applications of the methodology. Among them we can mention Avise (1992) on regional fauna; Hayes and Harrison (1992) on woodrats of the genus *Neotoma* of the eastern United States; Ellsworth and colleagues (1994) on white-tailed deer from the southeastern United States; da Silva and Patton (1998) on Echymid rodents of Amazonia; Walker and Avise (1998) on freshwater and terrestrial turtles of the southeastern United States; Aares and colleagues (2000) on two species of the genus *Phippsia* (Poaceae) in the North Atlantic region; Dawson (2001) on coastal marine animals from southwestern North America; Muss and colleagues (2001) on the reef fish genus *Ophioblennius;* and Fujii and colleagues (2002) on the plant genus *Fagus* from Japan.

Application of this methodology to the study of economically important pests is another development in this field. Among these studies we can cite Rozas and colleagues (1990) on the fruit-fly *Drosophila suboscura* in the United States; Chapco and colleagues (1992) on the North American grasshopper *Melanoplus sanguinipes;* Rich and colleagues (1995) about *Ixodes ricinus* in eastern North America; Brown and colleagues (1997) on *Greya politella* (Lepidoptera).

10

EXPERIMENTAL BIOGEOGRAPHY

ATTEMPTS to represent ecology and history as having separate, independent scales of evolution (for example, Brooks, 1988) are necessarily incomplete, since ecology and history play complementary roles in the generation of biogeographic patterns (Morrone, 1993c). In recognition of the integrated structure of ecology and history, theoretical evolutionary ecology is attempting to expand the spatial and temporal scales of ecology (for example, Cadle & Greene, 1994). On the other hand, Riddle and Honeycutt (1990) suggest that the explicit and rigorous formulation of historical biogeographical hypotheses have an increasingly important role in evolutionary biology, especially in the connections between macro- and microevolutionary phenomena. The connection between ecology and history requires new biogeographic models that relate this discipline to other sciences. Among the attempts, Haydon, Crother, and Pianka (1994) suggest that the biogeography of different taxa and regions may be conceptualized as a triangle involving complementary relationships among biology (the authors refer to ecology), history, and probability (Fig. 10.1). In other words, they visualize biogeography as an interaction among ecological, historical, and stochastic processes. Two of the

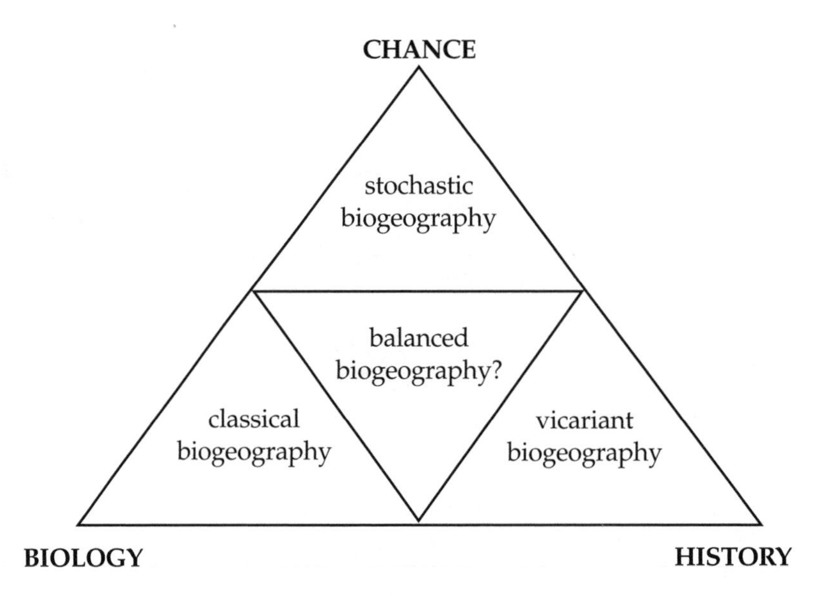

FIGURE 10.1. Experimental biogeography. The base of the triangle is the axis where historical factors increase in importance from left to right. On the left side, biology increases in importance toward the base. On the right side, the role of randomness (chance) increases in importance to the apex (modified from Haydon et al., 1994a).

three angles of this triangle are characterized by the different approaches of biogeography, namely ecological (called "classical biogeography") and historical (called erroneously "vicariant biogeography"). The other angle is characterized by what the authors call "stochastic biogeography." The interior of the triangle contains what they propose should be called "balanced biogeography," a synthesis among ecology, history, and stochastics.

A BIOGEOGRAPHIC MODEL

Haydon, Radtkey, and Pianka (1994) use this conceptual triangle to create a theoretical biogeographic model that explains biogeography in a hypothetical archipelago. They call this approach "experimental biogeography," corresponding to the previously mentioned "balanced biogeography."

Experimental biogeography uses a computer program (written in Pascal for a Sun 4 computer) to create a mathematical model in which vicariant events, phyletics, and ecology are perfectly known. The model exploits a hypothetical archipelago to explore interactions between ecological and historical processes and to determine their influence on local and regional diversity. The geological history is an independent variable overseeing the interaction of ecologically prescribed taxa with chance events. Mechanisms are developed that allow the evolution of faunal buildup to be replayed repeatedly over a fixed geological history. The hypothesized physical history of different biogeographic regions is programmed into the Sun 4, and taxa roam over the evolving substrate, exposed to certain predefined stochastic processes of dispersal, speciation, and extinction. Taxa are introduced that possess different (and nonevolving) vagilities and extinction-proneness, as well as other ecological attributes such as competitive ability. General questions on the model can then be posed, for instance whether radiations of taxa with different vagilities can be compared. Alternatively, specific questions can be addressed, such as assumptions about initial distributions of particular taxa; importance of directional variability in dispersal resulting from a current or prevailing wind; or the consequences of a proposed vicariant event. Through the use of such computer experiments, one can examine the sensitivity of the biogeographic pattern to assumptions made about generating processes. A simple way of applying this model simulates the fragmentation of one large island of area A_s over an arbitrary number of vicariant time units V_t. The probability of an island i splitting in any one vicariant time unit is a decreasing function, $V(a_i)$, of area a_i of the island. Upon splitting, the two new islands are allotted randomly selected speeds and directions, which are maintained until the occurrence of a subsequent split. It is thus possible to model the development of a random archipelago with a completely known geological history. Then, the evolution of a radiation within an intermediate-ordered taxon is simulated. The original island is seeded with one species representative of the taxon being modeled. Properties of taxa that are considered to remain constant within a radiation are vagility, propensity for interspecific competition, and proneness to extinction. Vagility, for instance, is modeled using a function, $D(d_{ij})$ that yields the probability of dispersal between islands i and j, separated by a distance d_{ij}. Every vicariant time unit is subdivided into an arbitrary number of "ecological" time units (E_t) in which each population of extant species on every island has a probability of dispersing from the islands on which they are found to all other islands, becoming extinct, and evolving niche position. When a population of a particular species becomes sufficiently isolated, the population is considered to undergo speciation. The model includes other variables that increase its complexity.

The main points of the model proposed by Haydon, Radtkey, and Pianka (1994) can be summarized as follows:

1. Ratios of vagility to vicariant spatial scales dictate ecological and biogeographic patterns.
2. The bulk of dispersal-speciation events occur within fairly well-defined windows of interisland distance; within these windows, reciprocal dispersal and speciation events generate diversity rapidly.
3. Archipelagoes may provide similar sequences of such windows for quite disparate taxa at different times, thus nurturing the proliferation of taxa with different vagilities at different stages in the geological development of the archipelago. The resulting biogeographic patterns may be remarkably similar.
4. Extinction-prone taxa proliferate species, many of which become extinct; non–extinction-prone taxa generate a higher extant diversity.
5. Proneness to extinction has a limited influence on beta diversity.
6. High extinction rates continually update the relationship between

the geological past and geographical distributions, eliminating the influence of "deep history," but influencing contemporary history.

7. Chance has its greatest influence on taxa with very low and very high vagilities (in other words, low-diversity taxa); its effect is enhanced at extremes of extinction proneness.
8. The imprint of history on the biogeography of taxa is very predictable.
9. Whereas alpha and gamma diversity may be quite unpredictable under some circumstances, their ratio (beta diversity) is generally very predictable.
10. Several measures of phylogenetic clade structure and development indicate that patterns within real clades are not likely to be adequately characterized by random structures or processes.

Experimental biogeography is a simulation model that attempts to reproduce within the virtual environment of the computer the structure and functioning of nature. It must be pointed out that the purpose of a simulation model is not to find answers about nature but to provide hypotheses about its structure and functioning. In other words, simulation models provide plausible hypotheses, not confirmations.

11

A COMPARISON OF METHODS: THE CASE OF THE SOUTHERN BEECHES

NOW THAT we have presented some of the primary methods of historical biogeography, it would be profitable to have a comparative view of some of them by means of a single empirical example—the plant genus *Nothofagus,* the southern beeches. It is not the aim of this chapter to evaluate or criticize methods but to compare some of them with the final objective of framing historical biogeographic hypotheses.

The study of the biota of the Southern Hemisphere has offered a wide variety of explanations for the patterns observed to overlay the complex history of Gondwana. Among the taxa that might explain these patterns is *Nothofagus,* which has traditionally been considered a "key genus" (Darlington, 1965; Van Steenis, 1971) for the study of terrestrial life in the Southern Hemisphere. *Nothofagus* (Nothofagaceae) includes 35 extant species divided into four subgenera (*Brassospora, Fuscospora, Lophozonia,* and *Nothofagus*) and is a characteristic element of cool-temperate forests in the circum-Pacific southern continents. The species of *Nothofagus* are distributed throughout southeast Australia, New Caledonia, New Guinea, New Zealand, Tasmania, and southwest South America (Fig.

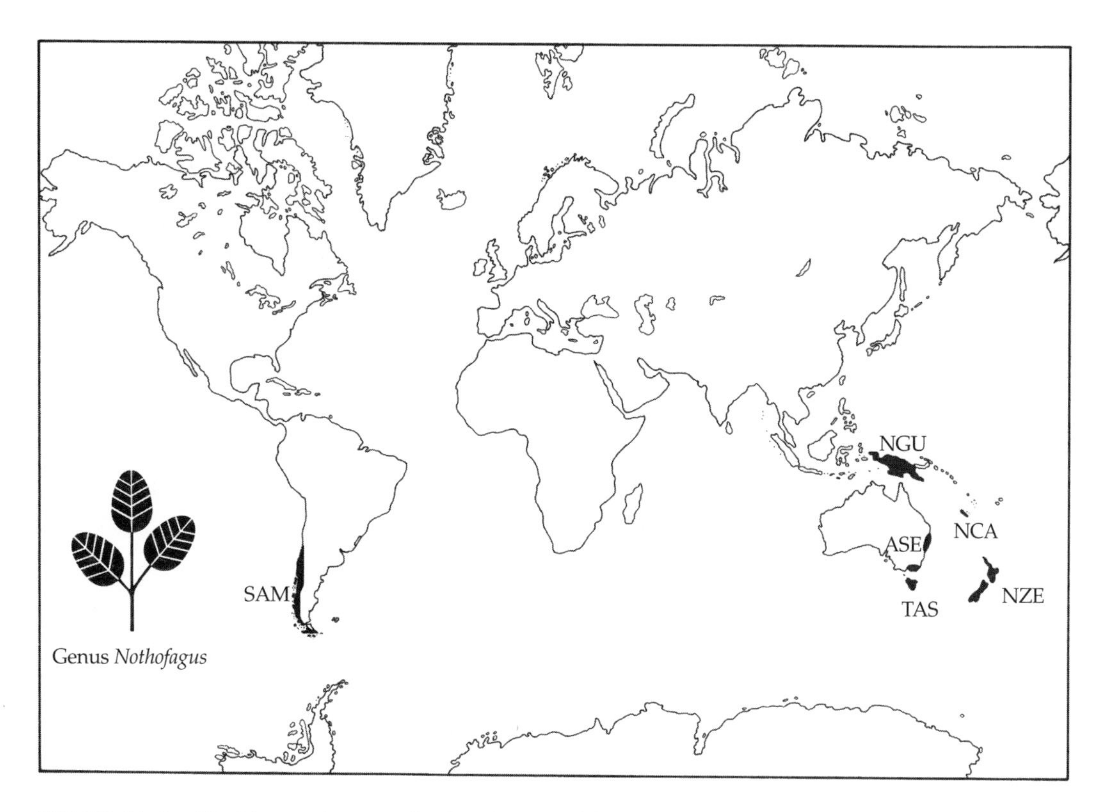

FIGURE 11.1. World map showing the distribution of the plant genus extant species of *Nothofagus*. ASE, southeast Australia; NCA, New Caledonia; NGU, New Guinea; NZE, New Zealand; SAM, South America; TAS, Tasmania.

11.1). The fruits of *Nothofagus* are single-seeded nuts, in groups of one to three, surrounded by a cupule. This type of fruit has a slow and restricted capacity of dispersal and its germination power decreases rapidly with age (Heywood, 1993). In addition, the seeds do not survive in sea water (Humphries, 1983). Such features of course affect the distribution of *Nothofagus* in the Southern Hemisphere. According to Hill (2001), the importance of *Nothofagus* for helping to provide general explanations of the taxa distribution in the Southern Hemisphere is mainly because it is usually prominent, grows on many of the southern landmasses, has an excellent fossil record, and its fruit is not adapted for wind dispersal.

The delimitation of areas of endemism, fundamental to historical biogeography (see chapter 1), would seem uncontroversial for *Nothofagus,* as most authors study the same areas. However, Nelson and Ladiges (1996, 2001) emphasize that one cause of paralogy is the imprecise characterization of geographic areas, and that areas commonly used to describe intercontinental relationships of *Nothofagus* (South America, New Zealand, Australia, New Caledonia, New Guinea, and Tasmania) are imprecise. Therefore, they suggest that a more precise characterization would expose some nodes as nonparalogous interrelating local areas within South America, New Zealand, and so on.

The literature about *Nothofagus* is quite extensive. Among the phylogenetic studies performed on this genus, we will mention Humphries (1981), Philipson and Philipson (1988), and Hill and Jordan (1993), whose studies are based on morphology; Martin and Dowd (1993), who use sequences of *rbc*L; Manos (1997), who use rDNA spacer sequences (ITS); Setoguchi and colleagues (1997), who use the *atpB-rbcL* intergenic spacer of the chloroplast DNA; Jordan and Hill (1999); and Manos (1997), who combines molecular and morphological data.

HYPOTHESES

Several contrasting hypotheses have been proposed concerning *Nothofagus*'s present distribution that involve dispersal, or vicariance and dispersal (see Humphries, 1981, 1983 and Nelson & Ladiges, 2001). Some of the invoked processes are:

1. Vicariance by breakup of Gondwana (Fig. 11.2) (Hooker, 1853; Engler, 1882; Croizat, 1952; Brundin, 1965; Humphries, 1981; Craw, 1989; Crisci et al., 1991a,b; Seberg, 1991; Weston & Crisp, 1994; Linder & Crisp, 1995; Ladiges et al., 1997; Nelson & Ladiges, 2001);

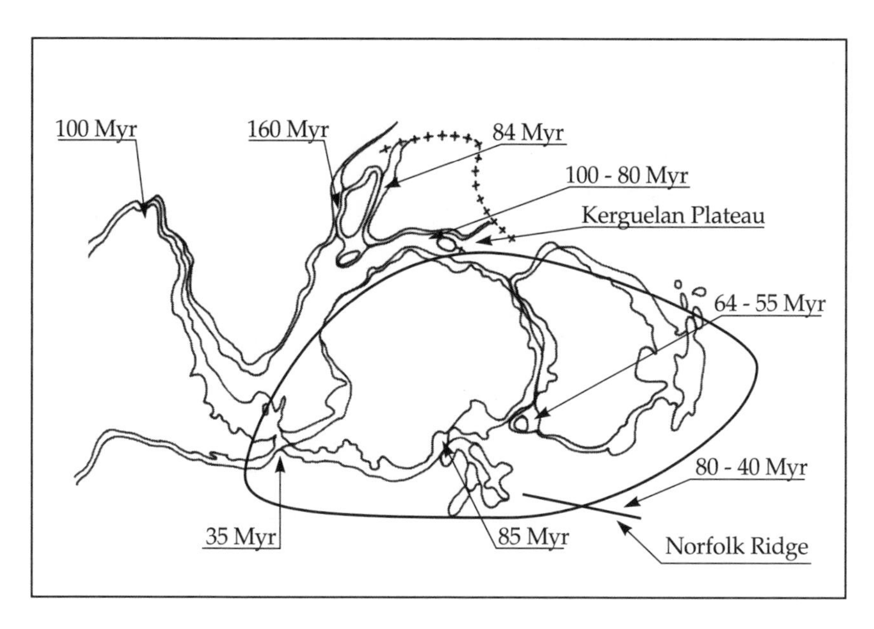

FIGURE 11.2. Hypothesis on the process and place of origin that led to the present distribution of the plant genus *Nothofagus* explained by vicariance from the breakup of Gondwana. Myr = million years ago.

2. origin in the Northern Hemisphere and migration by dispersal via Asia and/or South America to the Southern Hemisphere (Darwin, 1859; Wallace, 1876; Oliver, 1925; Schuster, 1976);
3. origin in Asia, crossing the tropics to Australia or New Zealand or both, radiation there, and a triple dispersal halfway around the southern end of the world (Darlington, 1965);
4. origin in Eurasia and dispersal via Africa-India (Fig. 11.3) (Raven & Axelrod, 1972, 1974);
5. origin in southeast Asia (Hill, 1992);
6. origin in South America plus Antarctica (Fig. 11.4) and dispersal (Hill, 1996; Hill & Dettman, 1996; Craw et al., 1999; Swenson et al., 2000, 2001);

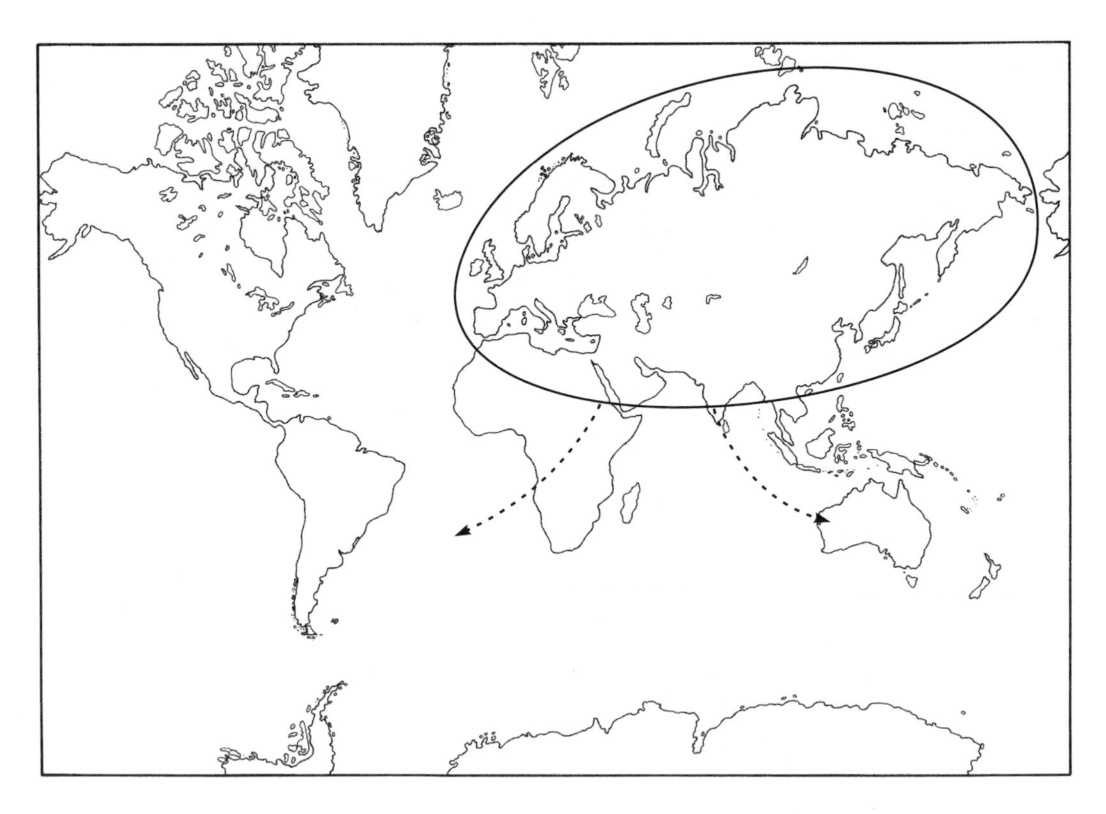

FIGURE 11.3. Hypothesis on the process and place of origin that led to the present distribution of the plant genus *Nothofagus* explained by an origin in Eurasia and dispersal via Africa-India.

7. origin in an area between New Zealand, Antarctica, and Australia and dispersal (Hanks & Fairbrothers, 1976);
8. origin in Chile and Patagonia and dispersal (Melville, 1973); and
9. origin in Antarctica and dispersal (Moore, 1972; Dettman, 1989).

The mode of dispersal was also subject to different hypotheses (Linder & Crisp, 1995); for instance, migration along land bridges (Florin, 1940; Couper, 1960; Van Steenis, 1962), long-distance dispersal for some or all of the species (Darlington, 1965; Pole, 1994; Hill, 1992; Martin &

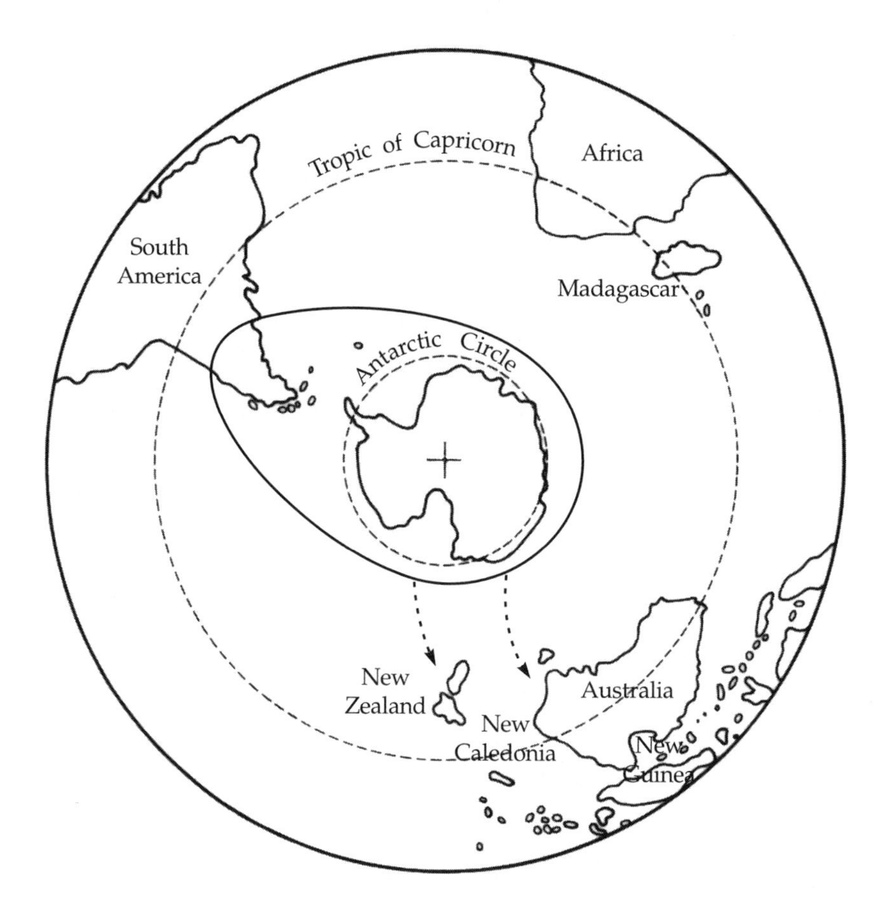

FIGURE 11.4. Hypothesis on the process and place of origin that led to the present distribution of the plant genus *Nothofagus* explained by an origin in South America plus Antarctica.

Dowd, 1993; Swenson et al., 2000, 2001; Swenson & Hill, 2001), or short-distance dispersal across much narrower ocean basins (Raven & Axelrod, 1974).

METHODS

The underlying hypotheses to explain the distributional patterns of *Nothofagus* and other taxa led to the application of different approaches and methods of historical biogeography. The approaches can be divided into two main types, those focusing on taxon biogeography, and those focusing on area biogeography.

Taxon biogeography aims to reconstruct only the distributional history of a taxon, in this case *Nothofagus.* Some examples are ancestral areas methodology as applied by Swenson and colleagues (2000), reconciled trees as applied by Swenson and Hill (2001) and Swenson and colleagues (2001), and dispersal-vicariance analysis (DIVA) applied by Posadas (unpublished).

Area biogeography aims to reconstruct the history of the areas, in this case the history of the areas inhabited by *Nothofagus* and by other taxa that share the same distribution. Applications of this are the cladistic biogeographic analyses of Humphries (1981), Crisci and colleagues (1991a,b), Seberg (1991), Ladiges and colleagues (1997), Humphries and Parenti (1999), and Nelson and Ladiges (2001); reconciled trees as applied by Linder and Crisp (1995); and panbiogeography as applied by Craw and colleagues (1999) and Katinas and colleagues (1999).

TAXON BIOGEOGRAPHY

Ancestral Area Analysis

Swenson and colleagues (2000) applied ancestral areas techniques (see chapter 4) to identify the most plausible ancestral area, or the region most closely affiliated with it, for *Nothofagus* by using Bremer's (1992), Ronquist's (1994), and Hausdorf's (1998) methodologies. The ancestral area analyses rest on Manos's (1997) cladogram of 22 extant species of *Nothofagus.*

Estimated values of the ancestral area analyses show that both Bremer's (1992) and Hausdorf's (1998) methods indicate that South America (the Weddellian province constituted by southern South America and the Antarctic Peninsula) is the most likely ancestral area. New Zealand is identified by both methods as the second most likely ancestral area, but the values are small compared to South America. Except for New Zealand and Tasmania, Ronquist's method gives an equally high estimation for all regions as being part of the ancestral area of *Nothofagus.* This method offers a less decisive result than Bremer's and Hausdorf's methods. Placing the ancestral area of *Nothofagus* in the Weddellian province is in accordance with the current fossil record, which suggests that southern South America and the Antarctic Peninsula played an important role in the initial differentiation and diversification of *Nothofagus.* The fossil record also provides strong evidence for secondary diversification of *Nothofagus* in southern Australia and New Zealand during the Palaeogene. The entire suite of *Nothofagus* species in New Zealand may have been lost at the time of marine transgressions in the early to mid-Cenozoic, and there is circumstantial evidence that *Nothofagus* was reintroduced into New Zealand from Australia during the Cenozoic via several long-distance dispersal events.

Reconciled Trees

Swenson and Hill (2001) investigated the biogeographical signal inherited in areagrams reduced from a well-supported phylogeny of *Nothofagus,* and attempted to determine whether, in a strict vicariance scenario, the areagrams predict all or part of the known fossil record. To do this they use two items, the *Nothofagus* phylogeny of Manos (1997), and the three most parsimonious areagrams derived from the taxon-distribution cladogram of Swenson and colleagues (2001). The biogeographic analysis was undertaken with the software COMPONENT 2.0 (Page, 1993). To convey information about extinct lineages, Antarctica was added to the areas inhabited by the extant species of *Nothofagus* as sister to South America.

Reconciled trees (see chapter 8) were produced between areagrams and the *Nothofagus* phylogeny following strict vicariance assumptions. The analyses identified six vicariance events and eight extinct lineages in different geographic areas. Known fossils of the genus throughout its present and past distribution range were optimized on the reconciled tree. The reconciled tree obtained to explain *Nothofagus* biogeography produced contrasting results among the subgenera. For the two subgenera that are relatively widespread today (*Lophozonia* and *Fuscospora*), the result ties in very closely with the fossil record. On the other hand, the subgenera *Nothofagus* and *Brassospora*, which have a relatively restricted present distribution (South America and New Guinea + New Caledonia, respectively), were widespread in the past, a conclusion not predicted by the areagram. Most parsimonious areagrams predicted one extinct lineage, possibly an unknown subgenus, formerly confined to New Zealand, southeast Australia, and Tasmania. This lineage has only poor support in the fossil record. Furthermore, the reconciled tree suggested six vicariant events, some of them between areas that are not geologically related. One vicariant event should have taken place between New Caledonia and New Guinea, although there is no geological support for the hypothesis that these two areas were closely related.

Dispersal-Vicariance Analysis

Posadas (unpublished) analyzed the distributional patterns of *Nothofagus* in the framework of event-based methods, applying dispersal-vicariance analysis (DIVA; see chapter 8).

To reconstruct the *Nothofagus* species ancestral distributions she used DIVA 1.1 (Ronquist, 1996), applying an exact search according to the dispersal-vicariance optimization proposed by Ronquist (1997b). The historical biogeography of *Nothofagus* was analyzed in terms of the phylogeny of the group proposed by Manos (1997) and modified by Swenson and colleagues (2001), who used 22 of the 35 extant species of *Nothofagus* as terminals. The six areas of endemism defined by Swenson and colleagues (2000) were used as units of the analysis.

According to DIVA there are eight alternative, equally optimal reconstructions that require four dispersal events. All possible ancestral distributions at each node are summarized in Figure 11.5.

Vicariant events. The highest frequency for vicariant events was assigned to South America, related to New Caledonia + New Guinea (22.22%). The other two major vicariant events had a frequency of 16.67

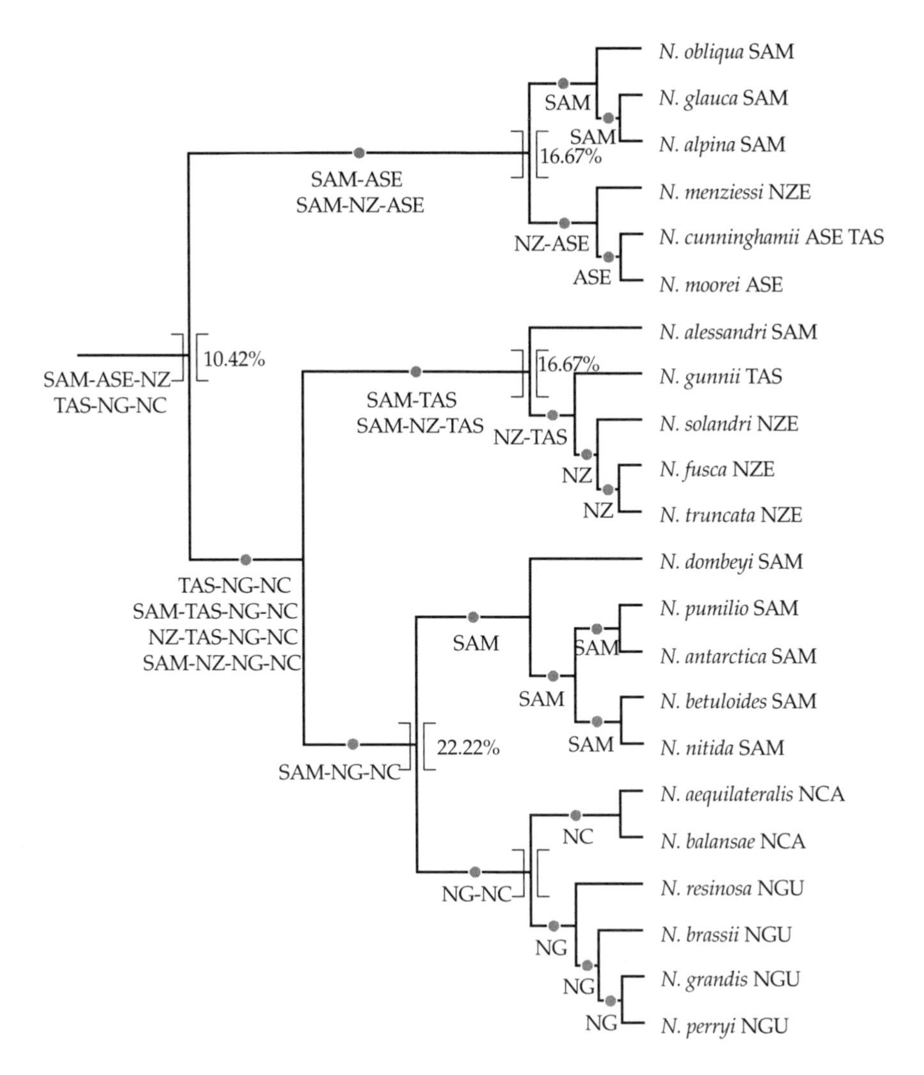

FIGURE 11.5. Application of DIVA in Manos's (1997) cladogram of 22 extant species of *Nothofagus* showing all alternative distributions at each node according to DIVA. Square brackets indicate vicariant events, their frequencies are indicated as a percentage. ASE, southeast Australia; NCA, New Caledonia; NGU, New Guinea; NZE, New Zealand; SAM, South America; TAS, Tasmania.

percent each. They implied that South America related to New Zealand + Tasmania, and South America related to New Zealand + southeast Australia. Thus, the three most frequent vicariant events involve the separation of South America from Australasian areas, which could reflect the breakup of Gondwana. There is still another vicariant event between New Guinea and New Caledonia, but as it involves only two areas it is not summarized by DIVA (DIVA summarizes only those vicariant events involving more than two areas). This vicariant event was also suggested by the results of the Swenson and Hill (2001) paper applying reconciled trees.

Dispersal events. The resulting dispersal events are all unidirectional. The highest score (57.14%) implied a one-way dispersal between two areas near the present location of southeast Australia and Tasmania. The remaining dispersal events were from southeast Australia to New Zealand (14.29%), Tasmania to New Zealand (14.29%), Tasmania to South America (10.74%), and southeast Australia to South America (3.6%).

AREA BIOGEOGRAPHY

Cladistic Biogeography

Several authors have attempted to provide a hypothesis for the austral global distributions in Gondwana. Do Gondwanan landmasses share a common ancestral biota to the exclusion of Laurasian landmasses? What are the cladistic interrelationships between different Gondwanan biotas? These are some of the questions that biogeographers are inclined to ask about the Southern Hemisphere (Hill & Weston, 2001), bearing in mind that geological or climatic events would produce vicariance of a once-continuous biota. Among those who have tried to answer these questions through the analysis of numerous groups of taxa, including *Nothofagus,* are Humphries (1981), Crisci and colleagues (1991a,b), Ladiges and colleagues (1997), and Nelson and Ladiges (2001) (see also chapter 6 on cladistic biogeography).

The study of Humphries (1981) was based on cladograms of *Notho-*

fagus and 24 other taxa; most of these cladograms were derived from previous cladistic taxonomic treatments. Humphries discerned two general cladistic patterns, using reduced area cladogram method, summarized as:

1. ((((Australia, New Guinea, Tasmania, New Zealand, New Caledonia) South America) Africa) (North America, Europe))
2. ((Australia, New Guinea) (South America, North America, Europe))

Humphries concluded that eastern North America and Europe are sister areas, and that there is a group of austral areas comprising New Zealand, Tasmania, Australia, New Caledonia, and New Guinea. The South American taxa either have close relatives in the Northern Hemisphere, or they have relatives in the Southern Hemisphere (Australia and associated areas), or relatives in other tropical areas such as tropical Africa (Humphries & Parenti, 1999).

Another attempt to resolve a general area cladogram for Gondwanan landmasses also found a subdivision in South America. Crisci and colleagues (1991a; see a full description of this example in chapter 6) analyzed 17 cladograms with primary Brooks parsimony analysis and component analysis and found that southern South America, New Zealand, Tasmania, Australia, New Caledonia, and New Guinea consistently grouped together as an austral biota, to the exclusion of northern South America, North America, and Africa. Interrelationships within the austral group, however, remain ambiguous.

It is interesting to note that differences in area delimitation (South America as a single unit versus South America as two units) may be the source of differences between the results of these two cladistic biogeographic studies.

Yet another application of cladistic biogeographic methods to study the history of areas inhabited by *Nothofagus* was made by Ladiges and colleagues (1997). These authors applied a paralogy-free subtree analysis, considering each of the 12 cladograms analyzed by Linder and Crisp (1995). According to Ladiges and colleagues (1997), for most cladograms there are repeated geographical areas, which is evidence of geographical paralogy. The cladograms of the plant genus *Aristotelia* and the fungus genus *Cyttaria* are the only two that potentially provide information on the relationship of South America, Australia, and New Zealand to one another. The two area cladograms were derived without the help of a computer once paralogy was recognized and subclades viewed separately as subtrees; the same results were derived using the TASS software, which, following input of taxon cladograms and listing of areas for each taxon, enumerates subtrees and prepares them as a matrix for analysis using a maximum parsimony program (Hennig86 in this case). The reduced area cladogram for *Nothofagus* based on the subtree analysis was: ((South America ((Australia, New Zealand) (New Caledonia, New Guinea)))).

Linder and Crisp (1995) demonstrated that *Cyttaria* relates Australia and New Zealand more closely to each other than to South America, but *Aristotelia* shows that South America is more closely related to New Zealand than to any other area. In Ladiges and colleagues' (1997) analysis, Australia and New Zealand are most closely related to each other than to any other area.

Reconciled Trees

Linder and Crisp (1995) assembled all plant taxa for which they could find reasonably robust phylogenetic hypotheses, and sought a parsimonious biogeographical pattern common to all. Phylogenies of *Nothofagus* together with 12 other plant taxa were taken into consideration to construct a general area cladogram. To locate a biogeographical signal in distributional data and phylogenetic hypotheses, the authors applied COMPONENT version 2.0 to map the area cladogram of each species onto the postulated general area cladograms and to measure the fit by a minimum number of items (duplications, deletions, or minimum number of independent losses or leaves) needed to reconcile the cladograms. Two analyses, based on assumptions 0 and 1, produced the same general area cladogram. The general pattern found for the relationship around the southern Pacific is: (South America (New Zealand (New Caledonia (Australia + Tasmania, New Guinea)))).

They used this general area cladogram in conjunction with the fossil record of *Nothofagus* to construct a historical scenario of the evolution of the genus. This scenario indicates extensive extinction, but also suggests that Australia has a more recent relationship to New Zealand than to southern South America. This is not congruent with the traditional geological theories, nor with the patterns evident from other sources such as insect biogeography. The results suggest general biogeographic congruence among the plants of eastern Gondwana, linking southern South America to New Zealand, Australia, New Caledonia, and New Guinea. This generality would suggest a single historical explanation, which is vicariance across a disintegrating Gondwana. The authors emphasize the vicariance hypothesis against the dispersal hypothesis.

Panbiogeography

The biogeographical affinities of South America, Australia, New Zealand, New Guinea, and New Caledonia, which share an austral biota, presumably would reflect a common history. However, as shown in the examples above, there are conflicting hypotheses of relationships concerning the austral areas, probably because historical biogeographic patterns reveal that there is more than one history of area relationships. In these cases the recognition of spatial homology, through a panbiogeographic approach (see also chapter 5), can be useful to discriminate among the different underlying histories. Panbiogeographic studies concerning the austral areas were developed using several taxa, including *Nothofagus*. We will refer only briefly to some of these studies, since they were cited or discussed in chapter 5.

Katinas and colleagues (1999) found an austral pattern represented by two generalized tracks with a Pacific baseline: One links southwest South America (the Subantarctic), Australia, and New Zealand; and the other links South America and New Zealand. The austral pattern was suggested to be related to Gondwana events in comparison with neotropical biotas. The individual track of *Nothofagus* linking South America with Australia, New Guinea, New Caledonia, and New Zealand do not form part of any of these two generalized tracks.

Craw (1985) and Craw and colleagues (1999) contrasted *Nothofagus* and the ratite birds and found that both taxa are not biogeographically homologous—each has different ocean baselines. *Nothofagus* is a member of a non-Gondwanan trans-Pacific fagalean alliance in the sense that its track, baseline, and main massings are not spatially congruent with lands bordering the Indian or the Atlantic Oceans, whereas the ratites are a Gondwanan group centered on Atlantic and Indian Ocean basins. According to Craw and colleagues (1999), Gondwana is understood to embrace taxa with distributions across the present Atlantic and Indian Oceans, but not across the Pacific. This is an interesting statement since the majority of the authors working with *Nothofagus* relate the genus to Gondwanan history. Craw and colleagues emphasize that *Nothofagus* is unknown as an authochtonous fossil from Africa and India, the heart of the Gondwana supercontinent.

FRAMING OF HISTORICAL BIOGEOGRAPHIC HYPOTHESES

A comparison of methods using *Nothofagus* has shown that different types of data (for example, a morphological taxa cladogram versus a molecular taxa cladogram), different assumptions (for example, dispersal versus dispersal and vicariance), and different methods (reconciled trees versus paralogy-free subtrees) often lead to different results. Also, different methods are more appropriate in different situations (for example, reconstruction of the distributional history of a taxon versus reconstruction of the history of areas of endemism). Direct tests of the methods are hampered by the lack of known biogeographical histories, when what is available are current records and mere traces of the past. However, the increased precision and rigor of the latest methodological developments in historical biogeography offer the possibility of removing biogeographical theories from a nontestable narrative status to one in which testable hypotheses (by additional phylogenetic, distributional, or geological data) of general significance may be generated.

These conclusions reflect the nature of the problem of framing historical biogeographic hypotheses. Research seldom fails because a researcher errs when collecting data or because of errors in the methodological calculations: Collecting data and making calculations are mainly objective. More often research fails because the researcher errs when subjectively deciding what data to collect and what method to use. The researcher errs by correctly solving, from the methodological point of view, the wrong problem.

Researchers must first subjectively structure and articulate their problem. Then they must objectively collect data and process it with their chosen method, solving the problem that they have articulated (Romesburg, 1984). It is with this in mind that we suggest the following framing for inquiries in historical biogeography:

1. *Identify a biogeographic problem and choose a research goal.* Define a gap of knowledge on historical biogeography, for example, the biogeographical history of *Nothofagus.* Create a specific question that will define the resarch goal, such as: What is the distributional history of the genus *Nothofagus?*
2. *Create a biogeographic hypothesis.* For example, the genus *Nothofagus* has been affected by dispersal, vicariance, and extinction, and these events are reflected in the geographical distribution and in the phylogeny of the genus. Furthermore, *Nothofagus* shares its biogeographic history with other taxa. The hypothesis should pass three tests (Ball, 1990): a) clarity (the units of study, areas, and taxa should be clearly defined); b) comparability (the hypothesis should be presented in such a way that direct comparison with hypotheses derived for other organisms in the same areas can be made); and c) rigour (the hypothesis should be presented so that potential observation, additional phylogenetic, distribu-

tional, or geological data that would serve to refute the hypotheses may be assessed.

3. *Choose a method.* The research goal and the hypothesis should influence the selection of a biogeographic approach. For example, the distributional history of *Nothofagus* is a taxon biogeographic question, therefore the chosen method should be among the ones that are consistent with this type of question. The choice should also reflect the hypothesis regarding the events that modify the geographical spatial arrangements of taxa. Therefore, in the example cited in steps 1 and 2, apply a method that contains vicariance, dispersal, and extinction, such as an event-based method. In short, choose the method that seems best suited to accomplish your objectives. Morrone and Crisci (1995) established that various methods (taxon biogeography and area biogeography methods) are not mutually exclusive—some of them can be integrated in a single biogeographic approach (see chapter 6). They can be applied at different levels of the analysis to resolve different problems.
4. *Apply the method to the data.* After a method is selected, collect and process data as specified by step 2 with the chosen method (on paper or computer). At this point it should be noted that there is often a strong temptation to substitute the means of research (step 4) for the ends of research (steps 1 and 2), to spend all of one's time choosing and using mathematical methods to process data while losing sight of one's purpose.
5. *Decide how the results inform and are useful.* Explain the results in terms of Earth history and evolutionary theory. Identify the conditions that would falsify or verify the explanatory premises and test the explanation. Finally, one biogeographic theory is better that another to the extent that it is more fruitful in suggesting scientific laws and more helpful in the formulation of explanatory hypotheses.

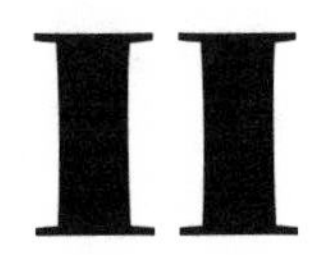

TOPICS IN HISTORICAL BIOGEOGRAPHY

12

MOLECULAR PHYLOGENIES IN BIOGEOGRAPHY

IN THE LAST DECADE molecular systematics has assumed an important role as an essential scientific discipline, both empirically rich and conceptually challenging. Molecules such as amino acids, proteins, RNA, DNA, and isozymes have been used to estimate the phylogenies of a wide variety of taxa. The revolution in gene sequencing technology has resulted in the production of more accurate phylogenies or gene genealogies that can be used to understand the biological processes occurring at many different levels of life's hierarchy. Genetics, development, behavior, epidemiology, ecology, conservation biology, and evolution represent fields that were illuminated by such information (Harvey & Nee, 1996).

As we mentioned in the introduction, historical biogeography was another field made visible by this newly opened window onto nature. Since 1994, there has been a great number of molecular systematics articles published in scientific journals that employ the word "biogeography" in the title or as an index term. These articles represent but a small portion of the total molecular systematics papers involving biogeography, since many studies deal with the topic but do not explicitly mention biogeography in the title. Most of these articles discuss molecular phylogenies in historical biogeography from two different perspectives, either

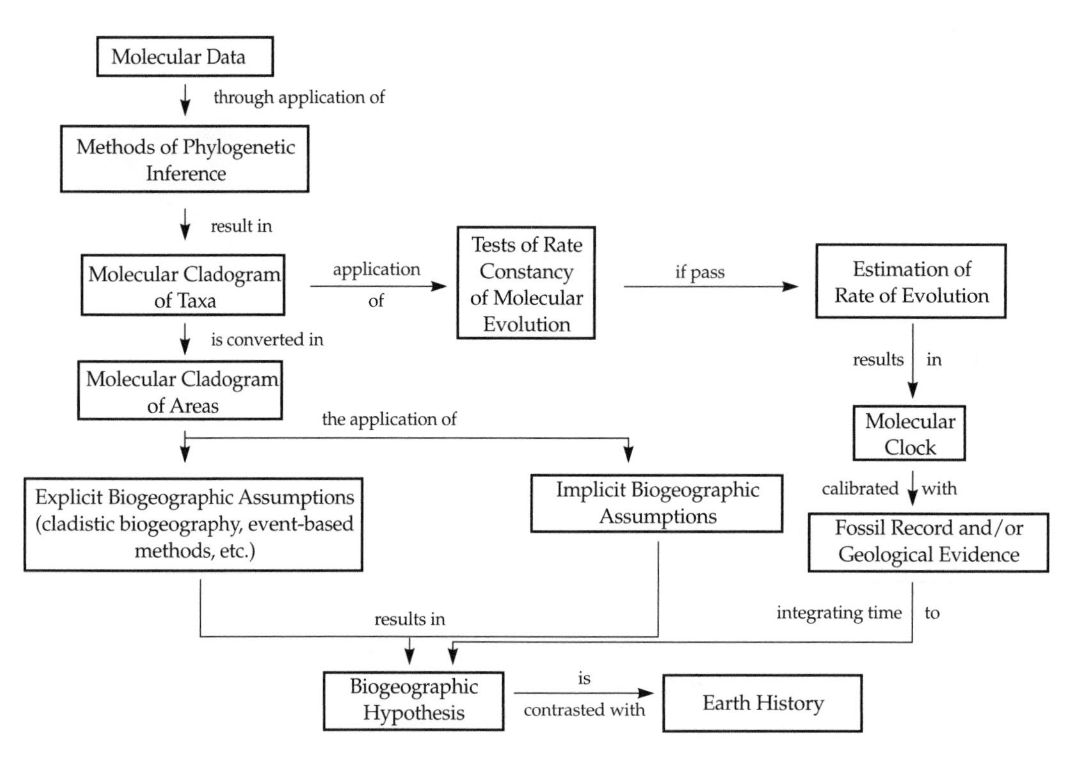

FIGURE 12.1. Flow chart of the role of molecules in historical biogeography. Recently, methods have been proposed that relax the stringency (rate constancy) of the clock assumption.

molecular cladograms as raw data of historical biogeography methods, or the molecular clock as a way to integrate time into the methods (Fig. 12.1). They constitute what was called "molecular biogeography," a term coined by Caccone and colleagues (1994) and retaken by Lavin and colleagues (2000) that attempts to reconstruct the biogeographic history of one taxon on the basis of its cladogram obtained from molecular data, with the additional application of the molecular clock.

MOLECULAR CLADOGRAMS

Cladograms obtained by applying a technique of phylogenetic reconstruction (see Appendix A) to a matrix of molecular data are used as raw data of historical biogeographic approaches such as phylogenetic biogeography (e.g., Knox & Palmer, 1998), reconciled trees (e.g., Swenson & Hill, 2001), paralogy-free subtrees (e.g., Brown et al., 2001), ancestral areas (e.g., Krzywinski et al., 2001; Swenson et al., 2000), and event-based methods (e.g., Beyra & Lavin, 1999; Donoghue et al., 2001). In these approaches, molecular- or morphological-based cladograms can be used as well, but other biogeographic approaches such as phylogeography demand the exclusive use of molecular phylogenies.

In some cases molecular cladograms are used in the context of a phylogenetic study to make further assumptions on dispersal and vicariance without an explicit application of a historical biogeographic method. It must be pointed out that the quality of the estimate of the biogeographic events depends upon the quality of the assumptions being made in the process of arriving at the estimate. It is very difficult, if not impossible, to arrive at an estimate of a biogeographic hypothesis without making assumptions, although it is quite easy to fail to realize the specific assumptions being made. For instance, many studies use the topology of the cladogram to explain a dispersal orientation of the group under study from the deeper branches to the top of the tree, using implicitly the assumptions of phylogenetic biogeography.

RELEVANCE OF TIME: THE MOLECULAR CLOCK

In a recent paper, Hunn and Upchurch (2001) emphasize the relevance of time when dealing with biogeographic problems. They argued that "data on the temporal distribution of taxa can provide an important additional constraint in biogeographical analyses. Such data may help to reinforce or overturn hypotheses of phylogenetic event causality." At the same time, they advocate a new paradigm for historical biogeography that they called the "chronobiogeographical paradigm." Hunn and Upchurch postulate that this change of paradigm in biogeography represents a logical elaboration rather than a replacement of the current paradigm. According to these authors, the successful implementation of a chronobiogeographical method has wide-ranging implications, and will allow us to reconstruct evolutionary histories and indicate missing data in both time and space simultaneously. The inclusion of such temporal information in biogeographic studies requires methodologies that allow us to assign time values to taxa, meaning the time of origination and the time of each cladogenetic event in a cladogram.

The rates of character evolution and the fossil record are two main sources of time information. The study of rates of character evolution has been a subject of interest in evolutionary biology since Simpson (1944), who, using information from the fossil record, concluded that the rates of evolution are highly variable. Some years later, Zuckerkandl and Pauling (1962, 1965) came to the opposite conclusion. In 1962 Zuckerkandl and Pauling proposed the theory of a molecular clock, stating that the rate of molecular evolution is approximately constant over time for all the proteins in all lineages. According to this theory, any time of divergence between proteins, genes, or lineages can be dated by measuring the number of changes between sequences (or proteins), since the molecular changes accumulate in populations in a clock-like fashion (that is, as a linear function of time). The difference between the sequences of a DNA segment in two species would then be proportional to the time since the two species diverged from a common ancestor (coalescence time). The "ticks" of the molecular clock correspond to substitutions or mutations. They do not occur at regular intervals as do the ticks of conventional clocks, but rather at random points in time (Gillespie, 1991). This time may be measured in arbitrary units and then it can be calibrated in millions of years for any given gene if the fossil record of that species exists. This hypothesis assumes that the gene under consideration is evolving neutrally, and that neutral mutation rates do not vary over time.

The rate of mutation (that is, clock speed) is assumed to be constant within a gene but variable among genes. For example, mtDNA has a mutation rate approximately ten times faster than the average chromosomally encoded human gene. The relatively rapid mutation rates of mtDNA sequences make them especially well suited to address more recent evolutionary events such as the relationships among the present human races. Other genes are more conservative and mutate less over time, such as the globin gene family. The genes that are the least likely to mutate are those coding for histones, cytochrome c, ATPases, and rRNA. For example, the rRNA genes are frequently used to infer relationships among major taxonomic groups including eubacteria and archaeobacteria.

As we mentioned previously, a molecular clock must be calibrated to obtain absolute rate estimates. This calibration is usually made by referring to the fossil record or to geological time (Sanderson, 1998). An example of the latter might be the date of the breakup of a continent (Hillis et al., 1996b). Calibrating the molecular clock by reference to geological history runs the risk of circular reasoning when the clock is used to test biogeographic hypotheses involving an event potentially caused by a geological process (for example, the breakup of the continents).

To calibrate the clock, first find at least two modern species for which the date of speciation can be determined from the fossil record, to establish the time since speciation, then determine the DNA sequence of the same gene in each of the two modern species, and infer or directly count the number of nucleotide substitutions between these two genes. All inferred or observed substitutions are assumed to have arisen subsequent to the putative speciation date. Therefore, the rate of DNA evolution for the gene under study is obtained by the number of DNA differences between the two modern species divided by the time since speciation. Assuming that the mutation rate for this gene is constant, we can then use the estimated rate to extrapolate the approximate dates of speciation for other species, for which we cannot determine a date of speciation from fossils. Calibration of the molecular clock may be difficult, because although taxa with a good fossil record can be used to calibrate a specific clock, the great majority of taxa lack a fossil record suitable for calibrating clocks and thus investigators must use a rate calibrated for other groups, a problematic but unavoidable approach (Voelker, 1999a).

The method described above works only if the gene under study is neutral with respect to selection. For this reason it becomes necessary to test the molecular clock hypothesis in the group under study. Testing the accuracy of the molecular clock has formed an important part of molecular systematics. Three tests have been proposed: the likelihood ratio test; the dispersion index; and the relative rates test (for a description of these tests, see Page & Holmes, 1998).

Phylogenetic analyses of DNA sequence data and the use of molecular clocks to estimate timing of genetic divergence can be used to test the biogeographic hypotheses. Depending on the relative ages of species divergence and vicariant events, assessments can be made of whether a dispersal hypothesis or vicariant hypothesis better explains the observed distributions. Clock calibrations that provide divergence estimates substantially smaller than those proposed by vicariant events suggest recent dispersal rather than ancient vicariance. Under a vicariance model, on the other hand, the phylogenetic relationships within clades, and associated divergence times derived from the molecular clock calibration, should be consistent with the order and timing of vicariant events (Waters et al., 2000).

More recently, the implementation of Bayesian methods has made it possible to estimate the error associated with tree topology, branch lengths, and nucleotide substitution parameters. In addition, Bayesian methods (see Appendix A) allow recognition of variation of the transition/transversion rate ratio to vary among sites, allowing the molecular clock to vary among lineages, and the use of codon-based or amino acid substitution models. In practice a Markov chain Monte Carlo algorithm is used to estimate the posterior distribution of the parameter values of interest (Huelsenbeck et al., 2000b; Huelsenbeck & Nielsen, 1999). An empirical application of this methodology can be found in Sequeira and Farrell (2001).

Numerous studies call into question the use of nucleotide mutations as a proxy time. These studies point out that mutation rates seem to vary both among and within genomes, being affected by many factors such as G (guanine) + C (cytosine) content (Wolfe, 1991), chromosomal position (Sharp et al., 1989), and nearest neighbor bases (Blake et al., 1992). Some earlier evidence suggests that mtDNA is subject to natural selection (Fos et al., 1990; MacRae & Anderson, 1988). Field evidence of lizards in islands of the Caribbean Sea (Malhotra & Thorpe, 1994) shows that instead of accumulating mutations steadily one at a time over millions of years, mutations in mtDNA can become rapidly fixed in a population, and major divergences in the mtDNA could have occurred in thousands instead of millions of years.

CASE STUDY: HISTORICAL BIOGEOGRAPHY OF THE COSMOPOLITAN PASSERINE GENUS *ANTHUS*

To assess the relative roles of dispersal and vicariance in the establishment of avifaunas, especially intercontinental avifaunas, a test for clock-like behavior in molecular data was applied, in conjunction with methods for inferring ancestral areas and dispersal events such as the ancestral areas (Bremer, 1992) and dispersal-vicariance (Ronquist, 1996) methods to a phylogeny rich in number of species, the cosmopolitan avian genus *Anthus* (Motacillidae) (Voelker, 1999a).

Defining areas: The areas employed were defined partly by previous authors in avian biogeographic analyses, and partly by *Anthus* breeding distributions. They involved areas in North America, South America, Eurasia, Africa, Australia, South Georgia Island, and the Canary and Madeira Island groups.

Obtaining the area cladogram: To assess the historical biogeography of *Anthus,* the maximum-likelihood phylogeny of the group obtained by Voelker (1999b) based on cytochrome *b* was converted to an area cladogram.

Methods: The dispersal-vicariance analysis (Ronquist, 1996) was used to reconstruct ancestral distributions on the phylogeny and the direction of dispersal events between areas. In addition, the ancestral areas method (Bremer, 1992) was used to provide an alternative to narrative dispersal from centers of origin scenarios. To test whether lineages within *Anthus* are evolving in a clocklike fashion, the two-cluster test (Takezaki et al.,

1995) was used. A 2 percent sequence divergence per million years in applying dates was used; this percentage has been inferred from several studies of disparate avian lineages.

Results: Despite the evidence that, overall, *Anthus* cytochrome *b* is not evolving in a clocklike manner, there are 25 of 40 nodes at which daughter lineages are evolving in a manner consistent with a molecular clock. The dates suggest that diversification of *Anthus* was high in the Pliocene (circa 7–2 million years ago) and low in the Pleistocene (2–0 million years ago); other avian groups show a similar pattern. The results of the ancestral area reconstructions by Bremer's method suggest several alternative possibilities for *Anthus* that involve Africa, Eurasia, and South America as probable ancestral areas. DIVA reconstruction suggests that either Africa or Eurasia are the most likely ancestral areas for the genus. Several other details suggest the likelihood of an eastern Asia origin of *Anthus* over any alternative area.

Anthus arose nearly 7 million years ago, probably in eastern Asia, and between 6 and 5 million years ago, *Anthus* species were present in Africa, the Palearctic, and North and South America. Speciation rates have been high throughout the Pliocene and quite low during the Pleistocene. Intercontinental movements since 5 million years ago have been few and largely restricted to interchange between Eurasia and Africa. Species swarms on North America, Africa, and Eurasia (but not South America or Australia) are the product of multiple invasions, rather than being solely the result of within-continent speciation. Dispersal has clearly played an important role in shaping the cosmopolitan distribution of this group. Molecular clock dates suggest that the two interchanges between South and North America predated the final uplift of the Panamanian land bridge. Island distributions resulted from dispersal. Furthermore, very limited distributions of several primarily Eurasian species in North America strongly suggest recent colonizations. Climatic shifts are the most likely vicariant events driving speciation between African and Eurasian forms. Vicariance may also be driving intracontinental speciation in *Anthus*.

RESEARCH USING MOLECULAR DATA IN BIOGEOGRAPHY

In the last few years many works have used molecular data in historical biogeography, as for example: Xiang and colleagues (1996) on the plant genus *Cornus;* Caccone and colleagues (1997) on the European genus *Euproctus;* Olmstead and Palmer (1997) on the plant genus *Solanum;* Xiang and colleagues (1998) on the Northern Hemisphere plant genus *Aesculus;* Morell and colleagues (2000) on the South American plant species *Gilia laciniata;* Waters and colleagues (2000) on the Gondwanic extant Galaxiid fishes; Chanderbali and colleagues (2001) on the Tropical and Subtropical plant family Lauraceae; Fritsch (2001) on the widely distributed plant genus *Styrax;* Hibbett (2001) on the Old and New World fungi genus *Lentinula;* Krzywinski and colleagues (2001) on the cosmopolitan insect subfamily Anophelinae; Renner and colleagues (2001) on the Pantropical plant family Melastomataceae. Recently, it has been an important trend on the study of island systems, where molecular information is applied to study colonization scenarios in such well-known archipelagos as Hawaii, Macaronesia, and southeast Asia. Examples of these studies are: Desalle (1995) on Hawaiian Drosophilidae (Insecta); Baldwin and Robichaux (1995) on Hawaiian silversword alliance (plant family Asteraceae); Juan and colleagues (1995) on Canarian darking beetles of the genus *Pimelia* (Insecta); Francisco-Ortega and colleagues (1997) on the Macaronesian genus *Argyranthemum* (plant family Asteraceae); Hahn and Systma (1999) on the southeast Asian genus *Caryota* (plant family Palmae); Helfgott and colleagues (2000) on the Macaronesian *Bencomia* alliance (plant family Rosaceae); Sun and colleagues (2001) on the plant genus *Helleborus;* and Davis and colleagues (2002) on the plant family Malpighiaceae.

13

BIODIVERSITY AND CONSERVATION EVALUATIONS

MANY PEOPLE worry about environmental problems—pollution, greenhouse warming, or ozone thinning—but do not take notice of the impoverishment of biodiversity (variety and variability of organisms and of the ecological systems that they constitute), a phenomenon whose very quick advance is creating an actual planetary crisis (Crisci et al., 1996). It has been calculated that at least half of the species that inhabit the planet will disappear during the next 50 years. This crisis of extinctions is comparable in its magnitude to the mass extinctions of the geological past, the last occurring 65 million years ago.

This change in global biodiversity is a complex response to several human-induced changes in the global environment. Sala and colleagues (2000) consider that the magnitude of this change is so large and so strongly linked to ecosystem processes and society's use of natural resources that loss of biodiversity is now considered an important global change in its own right. They identify five causes related to human activity that are the primary determinants of changes in biodiversity at the global scale: changes in land use, atmospheric carbon dioxide concentration, nitrogen deposition and acid rain, climate, and biotic exchanges.

New strategies need to be developed with the object of conservation. These strategies require biodiversity evaluations from the most possible perspectives. One of these perspectives ought to focus on where species live and have lived, and it is in this area that biogeography may participate in conservation evaluation. Biogeography as a discipline has the potential to aid our understanding and interpretation of the spatial patterns of biodiversity. Furthermore, as Grehan (1993) points out, "biogeography is the 'spatial model' for biodiversity that provides empirical evidence for a 'global biodiversity' structure in the real, natural world."

Biodiversity originated as a product of an evolutionary process that started together with life itself approximately 4 billion years ago. This evolutionary process has been strongly influenced by the geological history of our planet. Furthermore, understanding biodiversity requires knowledge of its history. Therefore, conservation of biodiversity unavoidable needs take into account historical information, both about the taxa themselves and about the history of the areas that they inhabit as well. Yet, in many cases biodiversity has been interpreted only in terms of numbers of species or other taxa, ignoring the historical and geographic factors. It is evident that the different regions of the planet have different biotic representations at the level of species as well as the level of higher taxa. This geographic variation of biodiversity is closely linked to history. Time and space are two factors which cannot be left aside when evaluating biodiversity (Brown & Lomolino, 1998).

BIOGEOGRAPHY AND CONSERVATION

Platnick (1992) affirms that one of the basic questions related to biodiversity conservation is actually a biogeographic question: How do we determine where the scarce economical and human resources should be invested to minimize the biotic impoverishment?

The development of methodological tools for biodiversity conservation requires a research program that includes a spatial dimension so that it is possible to understand how biodiversity varies in space and time

(Ulfstrand, 1992). This type of program is closely linked to the goals of biogeography. Biogeographic methods can be applied to biodiversity conservation—they are the perfect tools to outline the design of natural protected areas because they emphasize the spatial dimension of biodiversity and they contain geographical information (Craw et al., 1999). Prance (2000) agrees with the idea that biogeography has a major role to play in both the location and design of biodiversity reserves, but in his view it seems that only island biogeography has been adequately applied to the problem. The discipline can contribute much more, he claims.

Biogeography in conservation is generally funneled toward two questions. One is descriptive: the species distributional patterns, the identification of species distributional areas, the identification of areas of endemism, and the comparison of biotas corresponding to different areas. The second question involves analytical biogeography. Within this field, historical biogeography not only provides the necessary information for conservation (for example, recognition of spatial homologies, relationship determination between different areas), but also supplies methodological developments to be directly applied to determining priority areas for conservation (panbiogeography, PAE based on quadrats, phylogeography).

Biogeographic knowledge is absolutely critical for determining where conservation areas should be established (Humphries et al., 1995), especially when limited resources force us to prioritize. Species richness is the criterion most often used to select conservation areas, but this criterion by itself is not enough—comparison among the biotas (complementarity) is required. Biogeography provides useful information for the application of both the species richness and complementarity criteria, easing the identification of unique areas in terms of biotic composition. Platnick (1992) presents a clear example of this. Supposing that there are three areas to conserve and that each one hosts 5,000, 4,000, and 1,000 species, respectively (Fig. 13.1). The available financial resources permit the conservation of only two. A priori, it would seem that the most adequate solution would be to select the first two areas; yet, it could be that the

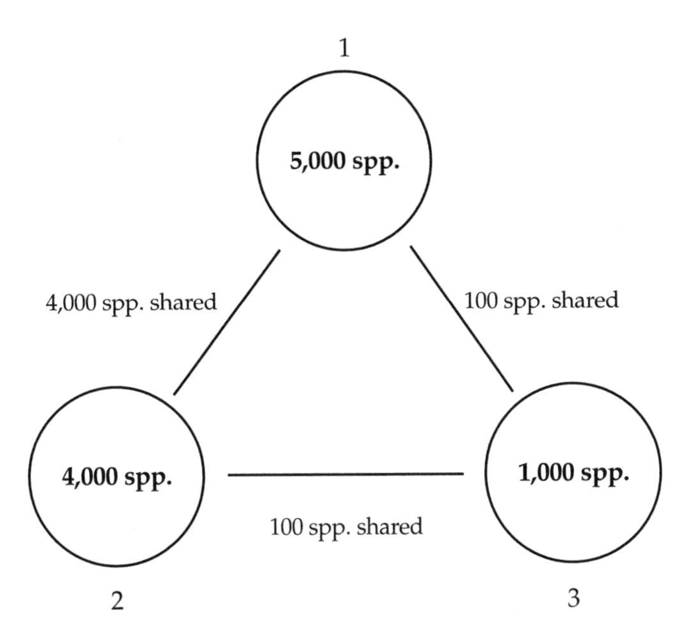

FIGURE 13.1. Platnick's (1992) example on the selection of conservation areas. If areas 1 and 2 were selected as priority areas for conservation (because of their richness), we would be protecting 5,000 species, because the 4,000 species present in area 2 are also present in area 1. If the choice takes into account the comparison of biotas, we would select areas 1 and 3, protecting a total of 5,900 species.

4,000 species present in the second area were also represented in the first, while of the 1,000 species present in the third area only 100 were represented in the first and second. In that way the selection of the first two areas would allow the conservation of 5,000 species in two areas, whereas the selection of the first and third areas would allow the conservation of 5,900 species in the same number of areas. Thus, it is possible to maximize the number of protected species and minimize the number of areas to conserve simply by comparing the biotas (complementarity principle).

But no single criterion could be used to adequately evaluate biodiversity conservation. It is necessary to integrate different approaches to give a broad perspective. In this sense, the evolutionary component of biodiversity is also important for evaluating priorities for conservation.

Barring major advances in molecular genetics, each time a species goes extinct the independent evolutionary features embodied in that lineage are lost forever (Heard & Mooers, 2000). Unfortunately, most conservation efforts are not using the historical dimension to its full potential. Cladograms are a powerful tool for summarizing the evolutionary history of life (see Appendix A). In the last decades several indexes of taxa phylogenetic information have been developed (Vane-Wright et al., 1991; Croizer, 1992, 1997; Faith, 1992a,b, 1993, 1994a,b,c). These indexes basically measure the taxonomic distinctness, or in other words, the "uniqueness" of the species. Underlying such indexes is the idea exposed by Heard and Mooers (2000) that phylogenetic information is one of the most important factors involved in any given level of extinction. It means that losses in evolutionary history will depend on the kind of diversification processes that gave rise to the clade under consideration (in particular, on the topology of the phylogenetic tree, which depends in turn on underlying variation in diversification rates across lineages within the clade).

Conservation planning requires indexes of biodiversity to be distinguished, labeled, and related to specific areas. How can we integrate in an index both the historical information of a taxa and biogeographic information (such as endemicity and complementarity) so that conservation priority areas may be ranked? The above-mentioned phylogenetic measures are calculated for each taxon in a cladogram, taking into account the history of the taxa. Thus, the index values could be used to produce scores for the areas as the sum of the index values calculated for each species that inhabit each area. Posadas and colleagues (2001) have added biogeographic information to these indexes. First, they use biogeographically defined areas as units (for example, regions, provinces, districts). Second, they modify the indexes to measure the degree of taxa endemicity. Third, they include complementarity to rank areas for conservation. To include a measure of endemicity, they propose splitting the index value of each individual taxon by the total number of areas that it inhabits. In this way, those taxa that are widespread will decrease in

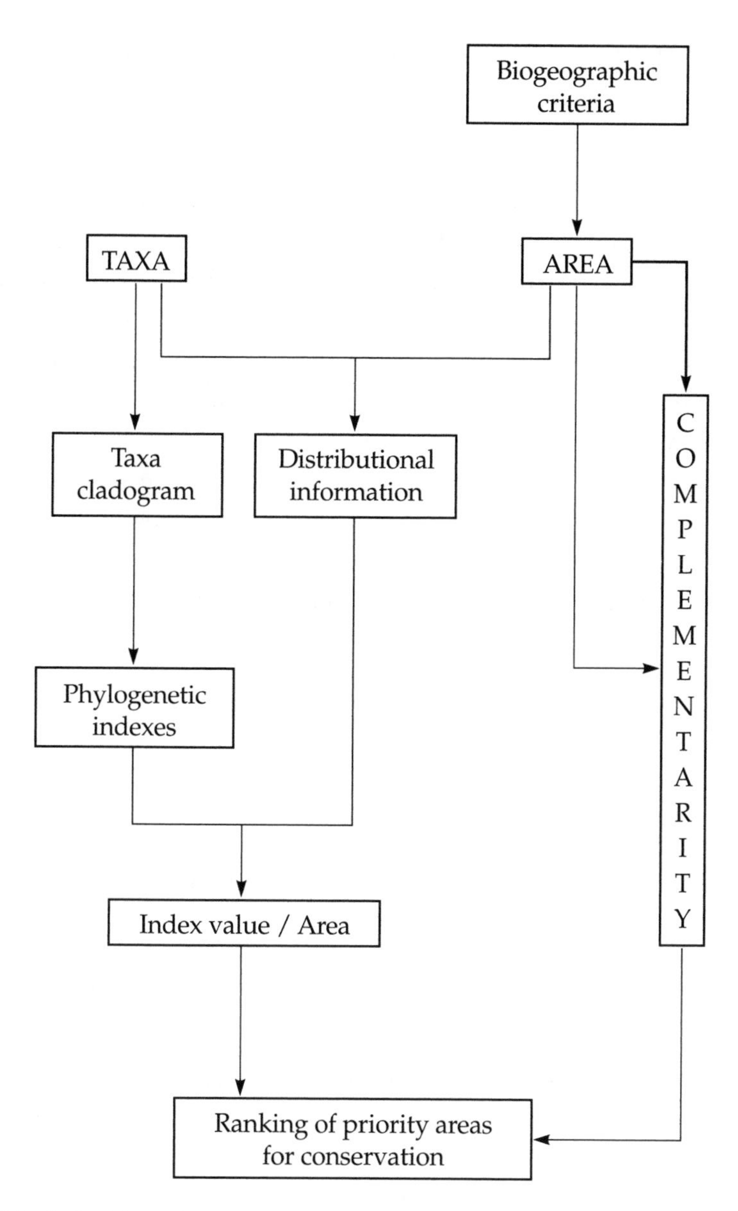

FIGURE 13.2. Flow chart indicating the steps proposed by Posadas and colleagues (2001) to rank areas for biodiversity conservation. This procedure attempts to integrate phylogenetic and biogeographic aspects of biodiversity when setting conservation priorities.

value in respect to those that are endemic to one area. To include complementarity in the evaluation, the highest priority area is determined by the highest index value (which accounts for taxonomic distinctness and endemicity of the taxa that inhabit it), the second area is selected based on complementarity (if more than one of the areas considered have maximum complementarity values related to the first area selected, then index values are used to choose among them), and so on, ranking all the areas under study.

Thus, this procedure (summarized in Fig. 13.2) allows us to combine information related to taxa evolutionary history and geographic information (such as endemicity and complementarity) in a biogeographic context (as the area units are defined using biogeographic criteria).

HISTORICAL BIOGEOGRAPHY METHODS APPLIED TO CONSERVATION EVALUATION

The concepts and methods of panbiogeography (see chapter 5) are applicable to local as well as global biodiversity problems. They provide useful criteria for the documentation, mapping, and recognition of the spatial characteristics natural to biodiversity (Craw et al., 1999). Such properties of panbiogeography concepts and methods has led Grehan (2000) to remark upon the advantage of establishing a global biodiversity atlas. Such an atlas would exploit the vast wealth of currently underutilized available distributional and systematic information. Furthermore, Miller (1994) identifies tracks and nodes as an approach for mapping biodiversity, which promotes the evolutionary infrastructure interpretation of the biosphere and aids in the identification of hot spots (biotic diversity centers). The importance of the historical biogeographic perspective is that it does not isolate biodiversity components from the space-time scenario. The concept of biogeographic homology sustained by the panbiogeographers introduces a new perspective to biodiversity studies, in which the space-time context is represented in the tracks, nodes, and baselines that identify different biogeographic centers and sectors (Craw et al.,

1999). The concept of a biogeographic node is of extreme importance because the nodes represent areas where organisms exhibit a local presence (endemism), a local absence (for example, absence of widespread groups in other areas), and diverse phylogenetic or geographic relations with other areas (Heads, 1990). The node represents points of confluence of two or more ancestral biotas widely separated in the past. Grehan (1993) emphasizes that nodes are biodiversity hot spots in a historical context, and consequently they must be considered as conservation priority areas. Recently, Morrone (2000) proposed the use of cladistic biogeography in addition to panbiogeography as a tool for conservation biology. A general area cladogram would help us to distinguish the basal clade of a threatened area, which could help restore the biota of that area.

The parsimony analysis of endemicity (PAE) based on quadrats (Morrone, 1994a; see chapter 7) gives a pattern of nested, hierarchically grouped areas (Morrone & Crisci, 1995), which can be used as an indicator of conservation priority areas. In this type of pattern the minor areas are included within progressively major areas. In this way, the smaller areas contain not only the unique species for the quadrats that represent them, but also those that are present in the major areas that contain them (Posadas, 1996). Cavieres and colleagues (2002) presented an empirical application of PAE based on quadrats specifically addressed to define areas of conservation priority.

Phylogeography and its methods (see chapter 9), according to Walker and Avise (1998), are relevant for biodiversity conservation at a genetic level. Knowledge of the phylogenetic and geographic structures of a population is fundamental for the conservation and management of endangered species. There is a view frequently expressed that populations and not species are the ultimate currency of biodiversity (Croizer, 1992; Mallet, 1995, 1996). From this view arises the concept of "evolutionary significant units" (ESUs). An ESU is one or a set of conspecific populations with a distinct, long-term evolutionary history mostly separate from other such units (Ryder, 1986). As such, ESUs are the primary sources of historical genetic diversity within a species (Moritz, 1995). As we pointed out in the previous section, it should be desirable to take into account historical information of taxa when setting priorities for conservation. The concept of ESUs provides a phylogenetic framework for deciding which population units are most distinctive, as they are the primary sources of historical genetic diversity within species (Moritz, 1995). But, how to identify an ESU? Phylogeography could provide the answer. Avise (2000) postulates that proper identification of ESUs within any species should rest on at least two facets of genealogical concordance (see chapter 9). Concordance across sequence characters within a gene merely identifies likely candidates for ESU status. Confirmation of such status then requires support from independent genetic characters within a species, spatial agreement in phylogenetic partitions across taxa, or congruence of spatial pattern with that from other independent biogeographic evidence. The phylogeographic recognition of ESUs is relevant for conservation at both the single species and regional biota levels. When working at the single species level it is evidently necessary to preserve the greatest amount of genetic diversity within the species. At a regional level, those areas within which multiple species display distinctive populations or ESUs deserve a higher status as conservation priorities.

Biodiversity conservation will be one of our most difficult challenges over the next few decades. That is why biodiversity evaluation must consider as many factors as possible. The phylogenetic measures adding biogeographic information, panbiogeography, PAE based on quadrats, and phylogeography generate high-value responses that reflect the taxa history and that of the areas that they inhabit (Crisci et al., 1999).

CASE STUDY 1: PANBIOGEOGRAPHIC NODES IN THE ANDEAN SUBREGION

A panbiogeographic analysis of the Andean subregion was performed to test the hypothesis of the hybrid origin of southern South America (Katinas et al., 1999; see also Crisci et al., 1991a). The details of this example were presented in chapter 5; perspectives from the point of view of conservation will be emphasized here.

Construction of individual tracks. Individual tracks were constructed for 154 taxa including fungus, plants, and animals.

Construction of generalized tracks. The superposition of the individual tracks brought about generalized tracks, which responded to three basic patterns that were named Andean, Austral, and Tropical.

Determination of panbiogeographic nodes and selection of conservation priority areas. The intersection of all generalized tracks gave three nodes as a result: Puna, Patagonia, and Subantarctic. These nodes represent complex biotic areas full of historical content that should be considered for defining priority areas for biodiversity conservation.

CASE STUDY 2: PAE IN TIERRA DEL FUEGO

The Tierra del Fuego archipelago is very interesting, not only because of its biota, but also because of the complex relations of this biota with those of other temperate areas of South America, Australia, New Zealand, and other minor islands of the austral hemisphere. With the objective of delimiting areas of endemism and determining areas for conservation in the Tierra del Fuego region, the parsimony analysis of endemicity (PAE) based on quadrats was applied (Posadas, 1996).

Delimiting areas. The Chilean and Argentine region of the Tierra del Fuego archipelago was considered, including major islands (Isla Grande) and minor islands (Isla de los Estados, Darwin, Navarino, Clarence, Desolación, Santa Inés, and Hoste).

Selecting taxa. The distribution of 377 species of vascular plants of Tierra del Fuego region was analyzed. The distributional data were taken from *Flora of Tierra del Fuego* (Moore, 1983).

Elaborating a grid. The Tierra del Fuego archipelago map was divided in 52 quadrats, most of them measuring a half-degree longitude and a half-degree latitude (Fig. 13.3).

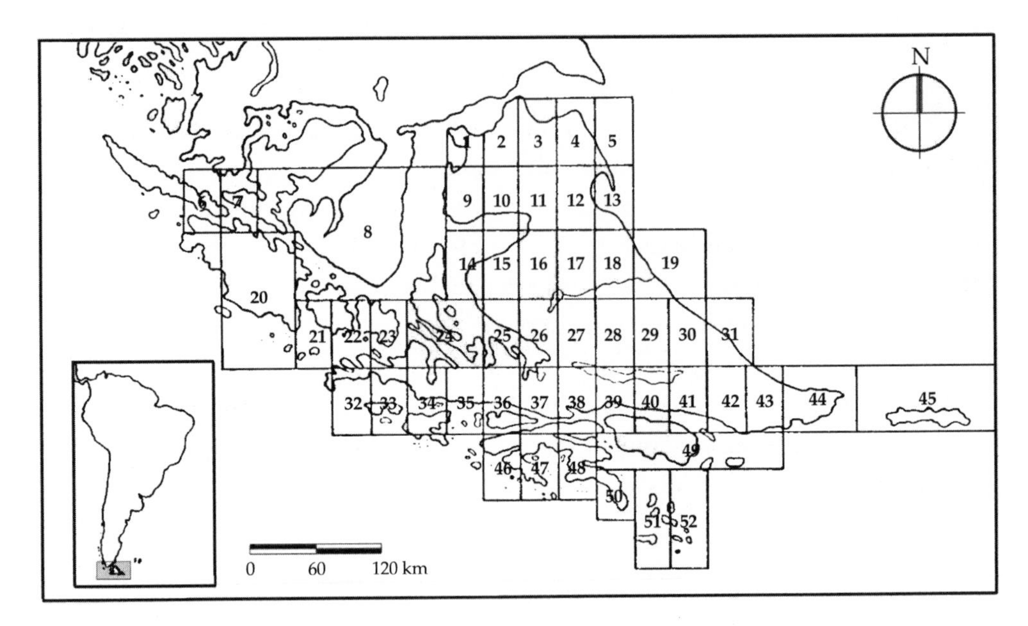

FIGURE 13.3. Tierra del Fuego archipelago grid with 52 quadrats to apply PAE based on quadrats (modified from Posadas, 1996).

Elaborating the data matrix. A matrix of 52 quadrats × 377 species was constructed. The presence of a species in an area was coded 1 and its absence 0. In the matrix a hypothetical quadrat all coded with 0 was included to root the cladogram.

Obtaining cladograms. The matrix was analyzed with Hennig86, options mh*, bb*, and the nelsen option were applied to obtain a strict consensus cladogram. Twenty most parsimonious trees were obtained (Length = 2,302; CI = 0.16; RI = 0.56), and a strict consensus tree was constructed (Fig. 13.4).

Delimiting areas of endemism. The strict consensus cladogram shows two major areas of endemism in the archipelago (A and B), while there are minor areas (C, D, E, and F) nested in the two former ones (Fig. 13.5).

Determining conservation areas. Five priority areas were determined

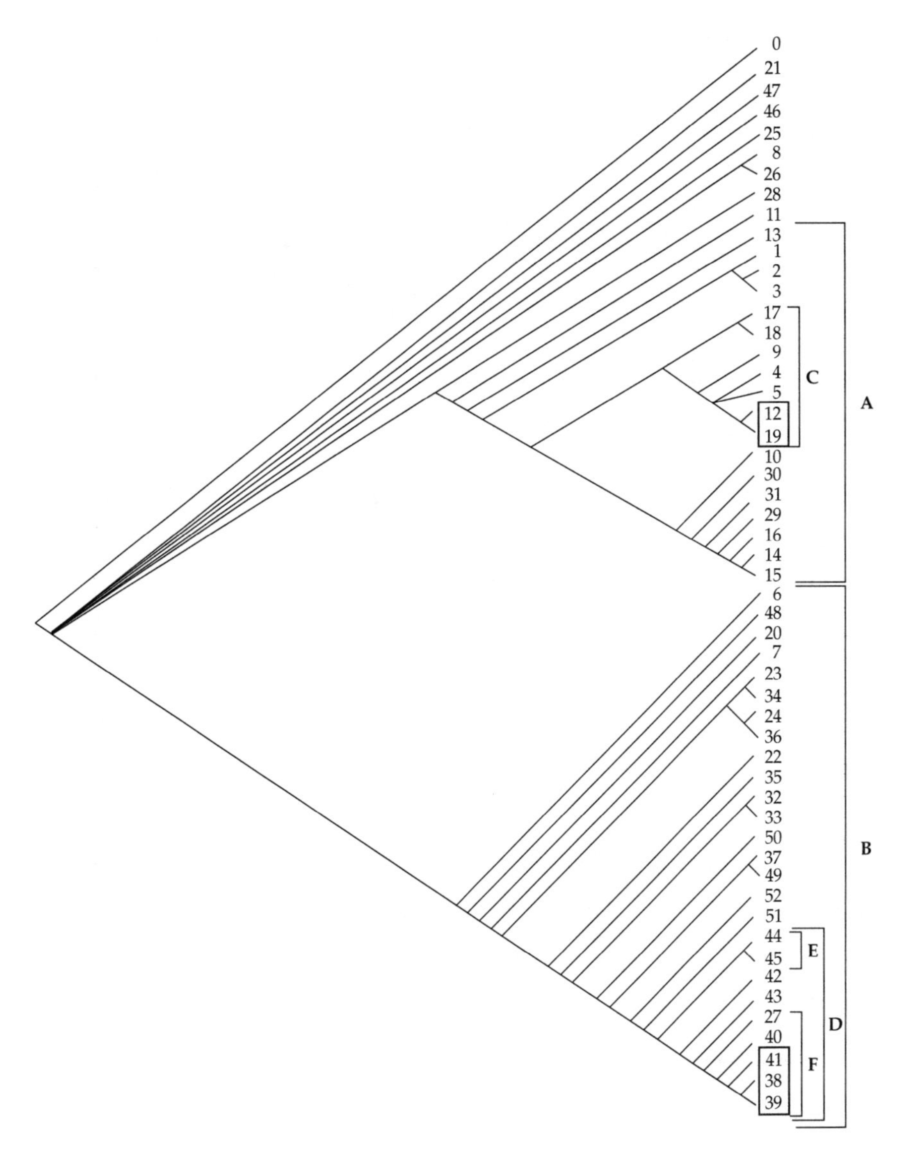

FIGURE 13.4. Strict consensus cladogram obtained from PAE. Each terminal represents one quadrat. A = Steppe; B = Magellanic Forest + Magellanic Moorland; C = Subarea within the Steppe; D = deciduous forest + southeast tip of Isla Grande + Isla de los Estados; E = southeast tip of Isla Grande + Isla de los Estados; F = Isla Grande deciduous forest. Priority areas for conservation marked with squares (modified from Posadas, 1996).

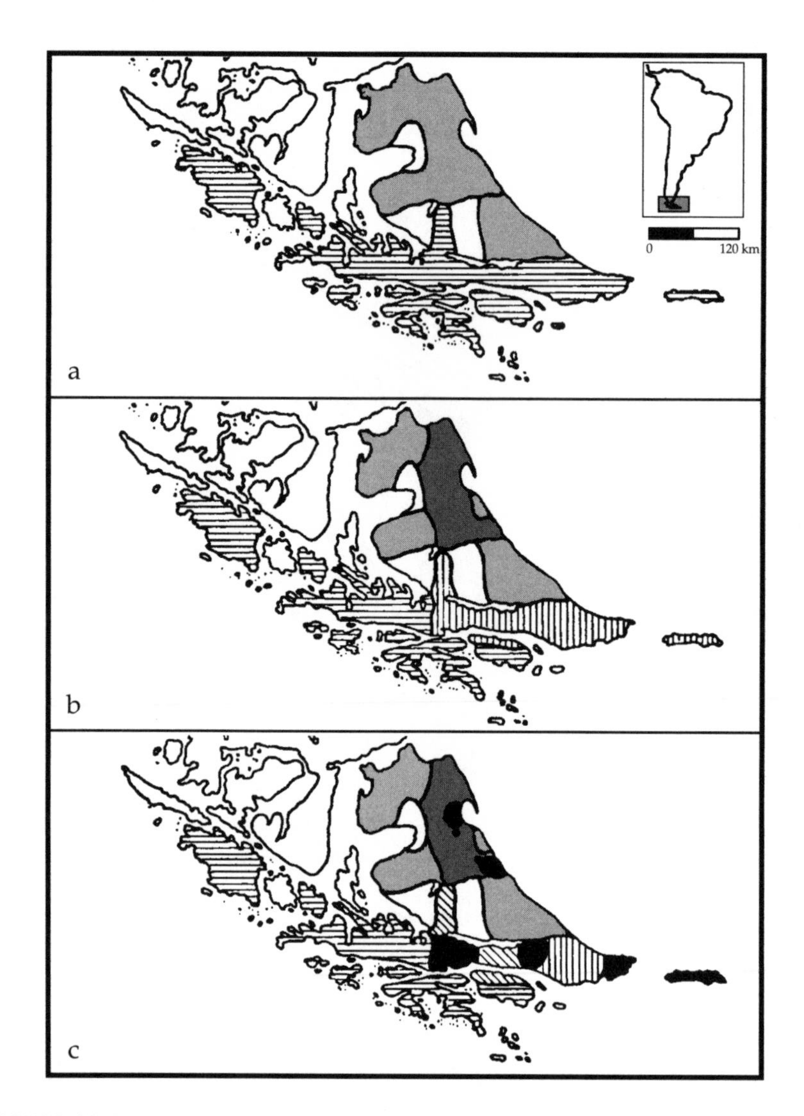

FIGURE 13.5. Tierra del Fuego maps showing the areas and subareas obtained by PAE. Light gray = Steppe; horizontal lines = Magellanic Forest + Magellanic Moorland; dark gray = subarea within Steppe; vertical lines = deciduous forest + southeast tip of Isla Grande + Isla de los Estados; oblique lines = quadrats 27, 38, 39, 40, and 41; black = priority areas for conservation (modified from Posadas, 1996).

for the conservation of biodiversity corresponding to the quadrats 12, 19, 44 + 45, 41, and 38 + 39, which are marked in black in Figure 13.5c.

From the point of view of biodiversity conservation, the nested area pattern obtained could be used for a quick and precise selection of priority areas to be conserved. The minor areas included in the nested area patterns that the cladogram shows contain all its species and also those present in the major areas in which they are included.

14

SPECIES INTRODUCTION

ALL ORGANISMS modify their environment, and humans are not an exception. Human enterprises transform the land surface, alter the major biogeochemical cycles, and add or remove species and genetically distinct populations throughout most of the Earth's ecosystem (Vitousek et al., 1997). These human activities result in global biodiversity change. Biotic exchanges (species introduction) have been considered as one of the most important drivers of such biodiversity change on a global scale (Vitousek et al., 1997; Sala et al., 2000).

Biotic exchanges from one continent to another constitute a natural biogeographic process. Classic examples are the rapid intercontinental expansion of the primitive horse or the waves of migration between the two Americas (di Castri, 1989; see chapter 2). But human dominance over the Earth's ecosystems has been accompanied by the widespread introduction of exotic species both deliberately and accidentally. The first human-induced introductions occurred as a result of the colonization and subsequent movements of primitive man. According to Brown (1989) fossil records of isolated oceanic islands are beginning to provide information on the surprisingly large impact of early humans on the biota. Early human colonists not only caused the extinctions of many native species, they also brought with them the first exotics. The growth of human population has shifted the once slow, erratic, and small-scale transfer of species to a rapid and large-scale translocation of large numbers and great species diversity (Ewel et al., 1999). Furthermore, the pattern of biotic exchange reflects the pattern of human activity. Remote areas with little human intervention receive fewer exotic species than areas that are in the middle of trade routes or that host intense human activity (Sala et al., 2000). The increasing rate of naturalization and spread (that is, rate of invasions) of species introduced both deliberately and accidentally poses an increasing global threat to native biodiversity, one ranked second only to habitat loss (Ewel et al., 1999).

SPECIES INTRODUCTION IS INDEED A BIOGEOGRAPHIC PROBLEM

Human-induced biotic exchange modifies the distributional patterns of species. As we suggested in the introduction to this book, historical biogeographers have identified three different space-time processes that can modify the geographic spatial arrangement of the organisms: extinction, dispersal, and vicariance. Two of these three space-time processes are identifiable in biotic exchanges. First, a dispersal event in biogeography implies that the range of the ancestral population was limited by a pre-existing barrier, which was crossed by some of its members. Then, they could colonize the new area. They may eventually differentiate into a new taxon if they remain isolated from the original population. Indeed, species introduction implies the breakdown of biogeographic barriers—introduction events could be assimilated to dispersal events. Second, extinction in biogeography implies the extirpation of a taxon from whole or part of its geographic range. There is wide evidence that species introduction, especially in oceanic islands, has led to the extinction of native species (Brown & Lomolino, 1998; Vander Zander et al., 1999).

BIOLOGICAL INVASIONS: SUCCESSFUL INTRODUCTIONS

Species introduction, as with any dispersal process, does not always result in colonization of the new area. Sala and colleagues (2000) remark that biological invasions (that is, successful establishment of exotic species) vary according to environmental conditions and biogeographic considerations. There is a wide variation within most biomes in the successful establishment of biotic introductions, depending on the original diversity and isolation of similar habitats. Thus, invasions have occurred least frequently in arctic and alpine ecosystems, because of their severe environment and the broad longitudinal distribution of much of the high-latitude flora and fauna; or in the continental areas of the tropics, because of high initial diversity and because abiotic and biotic factors characteristic of this bioma, including its high diversity, minimize the probability of successful establishment by invaders in undisturbed communities (Sala et al., 2000). Conversely, oceanic islands and southern temperate forests that have long been isolated are more prone to biotic invasions.

Brown (1989) postulates five rules for biological invasions of vertebrates:

1. Isolated environments with a low diversity of native species tend to be differentially susceptible to invasion: Those environments that are particularly susceptible to invasion by exotic species are generally those that are in some sense insular and have experienced low rates of natural colonization. They are usually characterized by small-size habitats that contrast markedly with the surrounding matrix, and by a long history of effective isolation from similar environments that would be the most likely source of suitable colonists. Examples include oceanic islands such as Hawaii and New Zealand, insular continents such as Australia and Madagascar, insular habitats such as lakes and desert springs, and other isolated distinctive environments such as the subtropical part of the Florida peninsula and the temperate forests of southern South

America. When immigration rates have historically been low, increasing the rate of colonization (in this case through human transport) is likely to result in the establishment of new species. Qualified in this way, this rule is simply a restatement of the theory of island biogeography (MacArthur & Wilson, 1967) and the extension of the theory to other kinds of insular habitats.

2. Species that are successful invaders tend to be native to continents and to extensive nonisolated habitats within continents: This rule is also a restatement of a long-held generalization of insular biogeography to the effect that there is a differential immigration from mainland to island.
3. Successful invasion is enhanced by similarity in the physical environment between the source and the target area: This pattern probably can be attributed primarily to the direct effects of physical factors in limiting the abundance and distribution of populations of vertebrates. In addition, availability of food resources and attributes of coexisting species tend to be correlated with physical variables. Consequently, colonists are more likely to be able to invade environments that provide climatic, geological, limnological, and oceanographic conditions similar to where they originated.
4. Invaders tend to be more successful when native species do not occupy similar niches: This rule is related to 1) and 2), and like them suggests that competition from native species with similar requirements plays a significant role in resisting invaders.
5. Species that inhabit disturbed environments and species with close associations to humans tend to be successful in invading man-modified habitats: Even on remote oceanic islands, many of the exotic vertebrates are associated with humans in their native regions, and their successful colonization must be attributed in part to their ability to exploit disturbed habitats.

Notwithstanding the fact that Brown (1989) established the aforesaid rules regarding vertebrates, they can be applied to most living beings.

INVASION CONSEQUENCES

Biotic invasions carry out changes in the natural environment and some of them are effectively irreversible. In some cases species introduction provides benefits for humans (crops, cattle, domestic animals), but in many cases deliberate and accidental introductions and subsequent biotic invasions have serious consequences for natural systems and humanity. Vitousek and colleagues (1997) present several examples of such consequences: Most infectious diseases are invaders over most of their range, degrading human and other species' health; some invaders become plagues and cause economic losses of billions of dollars. When affecting natural systems, invaders can disrupt ecosystem processes, altering the structure and functioning of whole ecosystems.

One of the most recognized consequences of biotic invasions is the loss of native species or populations (Vitousek et al., 1997; Brown & Lomolino, 1998; Ewel et al., 1999; Rahel, 2000). As Soulé (1991) postulated, the impact of a given factor depends on the time, the place, and the circumstances. In this sense, oceanic islands are dramatically susceptible to invasions. This susceptibility is clearly illustrated by two examples: On the Hawaiian archipelago the avifauna is currently composed of 95 bird species, 40 percent of which (37 species) are exotic, introduced in the last two centuries; in this span 14 native species have become extinct (Brown & Lomolino, 1998). All New Zealand terrestrial mammals (except two native species of bat) are invaders and have had an enormous impact on the native species of ground-nesting birds, which evolved in the complete absence of mammalian predators (Brown & Lomolino, 1998).

Perhaps the most important consequence of the biological invasions is an underappreciated aspect of global environmental change—the reduction of regional differences among floras and faunas, a process referred to as biotic homogenization (Rahel, 2000). Homogenization is a result of the loss of native species (many times determined by invaders) and the establishment of exotic species. Its effect is accelerated by the fact that the pool of introduced species is not a random sample of plants and animals. The information summarized by Ebenhard (1988) presents a clear example of that: Carnivores and even-toed ungulates constitute 19 percent and 31 percent of the introductions, respectively, whereas they represent less than 7 percent of the mainland species pool. Furthermore, Ebenhard postulated that species introductions of mammals and birds also had a strong biogeographic bias. On the continents, the number of species introductions appears to be roughly a function of the area of the region. Australia, however, appears to be an important exception to this species-area relationship, receiving many more introductions per area than any other continental landmass. Sixty percent of all mammal and bird introductions have occurred on oceanic islands. Introductions of plants, although of greater magnitude, were similar in their geographic bias. Before human colonization, over 90 percent of the flowering plants of the Hawaiian Islands were endemic. Now, nearly half of Hawaii's flowering plants are invaders. These biases and the loss of native species and populations increase the homogenizing effect of biotic invasions.

According to Ewel and colleagues (1999), areas of sociopolitical jurisdiction (states, countries, and trading blocks) are currently the units used for managing the movement of exotic species. Nevertheless, biogeographic barriers, uniqueness of local biotas, and dispersal capacities of exotic species do not necessarily mesh with political boundaries. As we postulate here, species introduction is a biogeographic problem and any attempt to control it requires a biogeographic context of analysis. Ewel and colleagues (1999) suggest that studies are needed to determine whether the current focus on political boundaries in regulating introductions produces substantially incorrect answers about their benefits and risks. Furthermore, they argue that perhaps a system such as that recently planned for Australia—in which natural ecological subdivisions, or bioregions, have been proposed to govern movement of exotic species—will prove to be most effective.

CONCLUSION: A CONCEPTUAL FRAMEWORK FOR THE FUTURE

THE PRECEDING sections have developed a schematic description of the present state of historical biogeography as far as it can go in this book. If this description has at all caught the essential structure of a discipline's revolution, it will simultaneously have posed a special problem. If a scientific revolution is a sign of a transition to a new paradigm (Kuhn, 1970), what needs to be done to generate the new foundation of historical biogeography—a foundation that implicitly will define the problems and methods of historical biogeography for succeeding generations of practitioners?

It is absolutely clear that the practice and philosophy of biogeography (as a whole, not only historical biogeography) depends upon the development of a coherent and comprehensive conceptual framework for handling the distribution of organisms and events in space-time. This task will require synthesis and novel conceptualization. The first and most important issue to be tackled is finding a solution for the ecology-history opposition. Ecology and history have always intermingled, they are indissolubly linked. Therefore, the long-established division between ecology and history is an obstacle to the progress of biogeography.

This necessary conceptual framework must also take into account the fact that biogeographical knowledge is fundamental to the study and management of environmental change. As Spellerberg and Sawyer (1999) note, biogeography holds the key to the survival of life. Future practical applications of this discipline may include:

Environmental monitoring to determine the effects on the biosphere of environmental changes;
public health management to measure the potential impacts of environmental change on human health;
management of commercially valuable species to identify the species most suitable for commercial enterprises in any particular location;
nature conservation to set priorities for biodiversity conservation; and
ecological studies and restoration to provide information about the former distribution of species and their respective communities and to guide development of ecological restoration initiatives.

The conceptual framework will require specifically from historical biogeography:

The development of rigorous tests for assessing the significance of biogeographic hypotheses (Morrone & Crisci, 1995);
more interaction with molecular biology;
more interaction with paleontology (for example, fossils as timing and palaeoecological indicators in biogeography; Pascual & Ortiz Jaureguizar, 1990);
a solution for the controversy generated by the proposition that organisms that are less capable of altering their geographic distribution are better historical subjects than the ones that possess powers of dispersal that allow them to alter their distributions much more readily (see Haydon, Crother, & Pianka, 1994; Templeton, 1998; Craw et al., 1999); and

the generation of high-quality empirical data (Morrone & Crisci, 1995). The scarcity of this kind of data hampers the progress of the discipline; Donn Rosen's Caribbean data from the 1970s has become legendary because almost everyone proposing a new technique still uses them.

Finally, the historical perspective provided by historical biogeography will continue to be a critical tool in the identification of sites worthy of protection for the conservation of species whose existence is threatened.

Biogeographers have made enormous strides in the past few years; in other words, we have started to babble in the language with which the traces of the past are telling us the history of life on Earth. But it is also in this precise moment that we discover the provisionality of our knowledge and that this history is so complex that probably we will never see it totally revealed.

Does this mean that our attempt to understand the distribution of organisms is a hopeless enterprise? Two voices answer this question. The first, Karl Popper (1959), is a voice of science:

> Science never pursues the illusory aim of making its answers final, or even probable. Its advance is, rather, towards the infinite yet attainable aim of ever discovering new, deeper, and more general problems, and of subjecting its ever tentative answers to ever renewed and ever more rigorous tests.
>
> The wrong view of science betrays itself in the craving to be right; for it is not his *possession* of knowledge, of irrefutable truth, that makes the man of science, but his persistent and recklessly critical *quest* for truth.

The second voice is that of art. It is not even a voice per se but a painting—*Sinbad the Sailor*—by the Swiss artist Paul Klee. It shows an oarsman in a small boat in a zone of lightness amidst a "sea" of darkness. There is little light, but enough to permit in rather slow motion the progress of the oarsman.

That is the way it is in any scientific enterprise, including biogeography. A journey like this is a difficult task that requires an attitude of modesty but not submission and, no doubt, it is a fascinating adventure. We certainly believe that in the perseverance of the oarsman is based his own glory.

APPENDIX **A**

PHYLOGENY

IN THE NINETEENTH CENTURY Lamarck, Cuvier, and Darwin recognized that life has a history. Darwin (1859), however, was the first scientist to suggest that a genealogical tree could represent this history. Müller published the first evolutionary tree in 1864. In 1866 Haeckel coined the word "phylogeny" to define the history of life. For almost a century after Haeckel, attempts to formulate a method for reconstructing phylogeny and its graphic representation as a tree failed or were incomplete. In 1950, the German entomologist Willi Hennig presented a basis for a method he called "phylogenetic systematics" (currently also known as "cladistics"). Cladistics, from the Greek word *klados* (branch), represented Haeckel's idea of a genealogical tree.

Cladistics aims to reconstruct the genealogies of organisms and to establish classifications that reflect those genealogies (Crisci & Katinas, 1997; Katinas & Crisci, 1999). The fundamental axiom of cladistics is that, as a product of evolution, nature is hierarchically ordered and this hierarchy can be represented by a branching diagram, called a tree or cladogram. This diagram is constructed on the basis of replicate sets of shared evolutionary novelties (synapomorphies) expressed in the most economical way (parsimony) (Crisci, 1982). Organisms can resemble each other because they share characters (attributes) that were present in a distant ancestor or because of characters that were present in the taxon that originated the group. For example, angiosperm taxa resemble each other because of the presence of transporting tissue, a character already present in the original ancestor of vascular plants or Tracheophytes. Angiosperms, however, also resemble each other by the presence of carpels constituting an ovary. This character, however, is only present in the most recent ancestor of angiosperms. So for angiosperms the transporting tissue is a primitive character, and the presence of an ovary is an advanced character (Fig. A.1). Thus, similarities among organisms can be hierarchically ordered because some characters appear earlier than others during evolution. Primitive ones are called plesiomorphous, and advanced or derived ones are called apomorphous. When these characters are shared by two or more taxa they are called symplesiomorphies and synapomorphies, respectively. Only synapomorphies indicate relationships of common ancestry among organisms, and all the organisms sharing a synapomorphy form a monophyletic group, which has its own history. A monophyletic group includes the common ancestor and all the species descending from it. For example, chlorophyll of plants and mammary glands of mammals constitute the synapomorphies that make plants and mammals monophyletic groups. The results of a cladistic analysis can be summarized in a branching diagram or cladogram. The branching points (nodes) of a cladogram show the order of appearance of the different apomorphic characters (Kitching et al., 1998).

There are several criteria that could be employed to infer when a character is derived or apomorphic (Crisci & Stuessy, 1980). The most reliable is through comparison of the group under study (ingroup) with a taxon or group of taxa closely related to the ingroup, called an outgroup. If a character occurs in the ingroup and in its outgroup, it is considered plesiomorphous because it was present in the common ancestor to both groups. But if it is absent in the outgroup and is unique to the ingroup, the character is apomorphous because it is considered an evolutionary

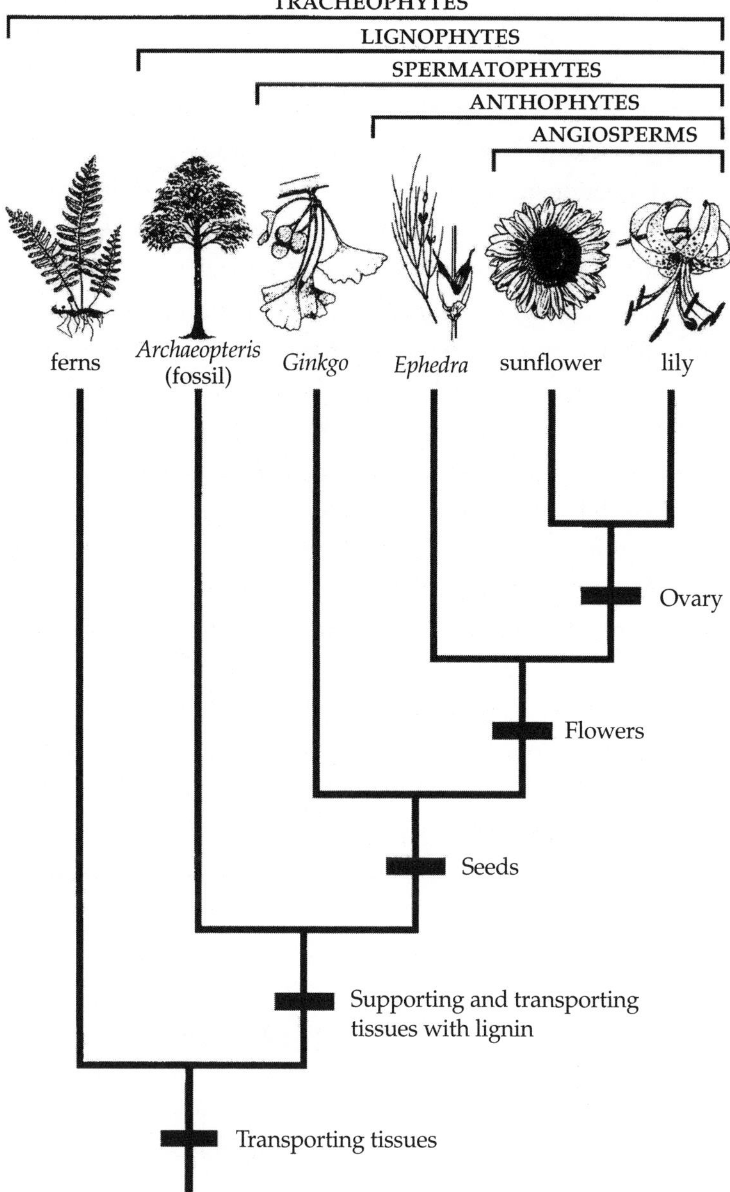

FIGURE A.1. Simplified cladogram of vascular plants. All vascular plants share an apomorphic character (transporting tissues). Apomorphic changes that define each monophyletic group are indicated on each node as bars.

novelty. Once that the synapomorphies are found, it is possible to construct on this basis numerous cladograms (hypotheses) that reflect the genealogical relationships among organisms. When choosing among cladograms, the explanation of character distributions that requires the smallest number of changes, meaning the shortest tree(s), is preferable (principle of parsimony). The number of trees that must be evaluated under the parsimony principle increases exponentially with the number of taxa included in the study. Kluge and Farris (1969) developed the so-called Wagner parsimony technique, which is a quantitative way of doing cladistics. Through the years this technique has been implemented using different algorithms in various computer programs. The original technique proposed by Kluge and Farris (1969) underwent many changes to improve it. All these subsequent algorithms can be grouped under the title of parsimony or maximum parsimony algorithms. Also variants of parsimony have been proposed, such as Fitch parsimony, which treats multistate characters as nonadditive or unordered; Dollo parsimony for data, in which each evolutionary novelty is assumed to have arisen only once on the cladogram (parallel gains and loss regains are prohibited); and Camin-Sokal parsimony, which postulates that character evolution is irreversible and all homoplasy must be accounted for by parallelism or convergence.

There are two main types of characters that can be used when performing a cladistic analysis, morphological and molecular. Morphological characters include external morphology, internal morphology or anatomy, embryology, palynology, and so on. Molecular characters or molecular markers, on the other hand, may include deoxyribonucleic acid (DNA) sequencing, protein electrophoresis, immunological assays, DNA-DNA hybridization, restriction fragment length polymorphism (RFLP), randomly amplified DNA polymorphisms (RAPD) and similar methods based on PCR (polymerase chain reaction), and protein sequencing.

The utility of each of these types of molecular markers depends upon the evolutionary or systematic question being considered (Avise, 1994).

Although nucleic acid sequencing is a comparatively new approach for systematics, the power of the technique has ensured that DNA sequencing has become one of the most utilized of the molecular approaches for inferring phylogenetic history. One of the main advantages DNA sequence data has over other kinds of molecular markers is that comparison of putatively homologous genes from different taxa can be made directly from multiple sequence alignments (Crisci, 1998c). By means of this method, it is possible to know the order (sequence) alignment of the four nucleotides in a fragment of DNA. With the sequence alignments information of several taxa it is possible to construct a cladogram applying a maximum parsimony algorithm, following the same steps as in a morphology-based cladistic analysis. With both kinds of data, morphological or molecular, the steps for constructing a maximum parsimony cladogram are as follows:

1. Construct a taxa × characters matrix, in which the characters (morphological and/or molecular) are codified.
2. Include an outgroup (OG) to establish character polarity, i.e., to determine the primitive-state characters (this step is not necessary when you are constructing an unrooted cladogram).
3. Search for the cladogram that requires the fewest ad hoc assumptions to explain the distribution of character states (that is, the shortest tree/s). Figure A.2 shows an example of the application of the method to molecular data.

Methods other than parsimony have been developed for the purpose of phylogenetic reconstruction. Some authors (for example, Edwards & Cavalli-Sforza, 1964; Cavalli-Sforza & Edwards, 1967; Felsenstein, 1985; among others) have proposed that phylogeny estimation should be viewed as a statistical problem, and that cladogram construction should be treated as a problem of statistical inference (see Edwards, 1996 and Felsenstein, 2001 for a discussion of the origin, growth, and spread of statistical phylogenetics). The likelihood models assume that evolution is a

Sites	Nucleotides
	1 2 3 4
OG	A A T T
Taxon a	G C T A
Taxon b	G C T A
Taxon c	G A C A
Taxon d	G A C A
Taxon e	G A T T

a

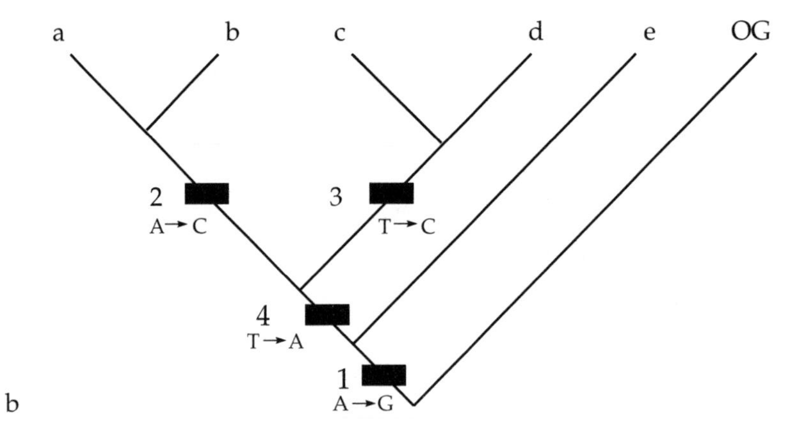

FIGURE A.2. Cladistic analysis applied to molecular data. *(a)* Data matrix where a nucleotide sequence of a DNA region is compared (constituted by four sites: 1, 2, 3, and 4) of five taxa (a, b, c, d, and e) and one taxon used as an outgroup to root the cladogram. A = adenine, C = cytosine, G = guanine, T = timine, OG = outgroup; *(b)* cladogram obtained from data matrix showing the relationships among the taxa a, b, c, d, and e. According to the cladogram, a and b constitute a closely related group, and c and d constitute another closely related group. Arrows represent nucleotide changes from a primitive state to a derived one, and are marked in the node where the change appears.

stochastic process for which each possible phylogeny can be assigned a probability given a particular set of data. Thus, the preferred tree is the one that is most likely to have given rise to that particular data set, which may or may not be congruent with the tree as inferred by the parsimony criterion. Among these approaches to phylogeny reconstruction can be mentioned the maximum likelihood methods (Felsenstein, 1981; Swofford et al., 1996). Most applications of maximum likelihood have dealt directly with nucleotide sequence data. More recently, however, it has been suggested that maximum likelihood methods be applied to discrete morphological characters (Tuffley & Steel, 1997; Lewis, 2001a,b).

The term maximum likelihood does not refer to a single statistical method, but rather to a general approach of inferring phylogenetic relationships using a prespecified model of sequence evolution. Maximum likelihood favors the tree topology that is most likely for a specific model of nucleotide substitution and branch lengths, given the observed sequence data. Just as with maximum parsimony, the number of trees that must be evaluated under the likelihood criterion increase exponentially with the number of sequences included in an analysis.

The objective of maximum parsimony is to minimize the number of substitutions on a tree topology. This strategy seems reasonable if we are willing to assume that all branch lengths on a given topology are equal. In reality, however, branch lengths are probably variable in length. Therefore, when employing an optimality criterion that includes branch length information in the calculation of the objective function, one might find a more likely tree topology even though some of the sequences share a common character but do not share a common ancestor if the branches separating these sequences are relatively long (Swofford et al., 2001).

The general steps to calculating the likelihood of a given tree topology are as follows:

1. Calculate the likelihood of the tree for each possible combination of ancestral character states given the observed state of the terminal characters.

2. Sum the likelihood scores for each set of ancestral character states.
3. Repeat steps 1 and 2 for each nucleotide site in the alignment.
4. Multiply the likelihood scores for each nucleotide site to get the total likelihood of the tree given the data. In practice, the likelihood scores are represented as logs and therefore the scores for each site are summed rather than multiplied.

Likelihood assumes that the evolution of each site is independent, and therefore, the product of the likelihoods for each site provides the overall likelihood of the observed data. Finally, different tree topologies are evaluated to find the best one. An overview of the calculation of the likelihood of a tree is presented by Swofford et al. (1996: 431, fig. 10) in a simple example of four taxa.

Bayesian methods are among the most recent developments in phylogenetic analysis (Rannala & Yang, 1996; Mau & Newton, 1997; Larget & Simon, 1999; Huelsenbeck et al., 2001). As described so far, the maximum likelihood method estimates the parameter values (for example, phylogeny, branch lengths, and nucleotide substitution model) that maximize the likelihood function. Bayesian inference of phylogeny is based upon the posterior probability of the parameters, which is obtained from the likelihood function and the prior probability of the parameters. Because a number of parameters must be estimated simultaneously, simulations are used to obtain a sample from the posterior distribution and to base inferences on this sample. More specifically, a Markov chain Monte Carlo (MCMC) algorithm is used to solve the computational aspects of the problem of sampling phylogenies according to their posterior probabilities of trees (Larget & Simon, 1999; Yang & Rannala, 1997). A Bayesian approach to phylogeny reconstruction requires a likelihood model for sequence evolution through a phylogenetic tree, prior distribution on trees and model parameters, and data (Hall, 2001).

Finally, it must be pointed out that other phylogenetic reconstruction methods have been proposed, such as distance matrix methods, in which evolutionary distance (for example, numbers of nucleotide substitutions between sequences) are computed for all pairs of taxa, and a phylogenetic tree is constructed by using a technique based on some functional relationships among the distance values (Swofford et al., 1996). Unlike maximum parsimony, maximum likelihood, and Bayesian methods, which are character-based approaches (they all use the taxa × characters matrix), distance methods use a distance matrix to reconstruct the phylogeny.

APPENDIX **B**

SOFTWARE IN HISTORICAL BIOGEOGRAPHY

HISTORICAL BIOGEOGRAPHY has not been left behind the other natural sciences in employing software for research and analysis; on the contrary, such software acted as a catalyst in the theoretical and practical transformation of the subject (Crisci et al., 1994). In the last two decades of the twentieth century, an abundance of software was produced to reconstruct the spatial history of life on Earth.

Software useful in historical biogeographical research and analysis is listed in the order of the methods presented in chapters 2–10, except for the programs that apply phylogenetic inference algorithms (maximum parsimony, maximum likelihood, and Bayesian estimation), which are listed independently. The phylogenetic programs are applicable to several of the approaches for elaborating taxa cladograms, which are the raw materials of most of the biogeographic methods.

In historical biogeography, the selection of the method determines (in most cases) the software to be used, as each program has been developed for a certain approach, and each approach generally has only one available program.

The extraordinarily rapid development in both software and hardware make this appendix an inevitable anachronism. Yet we think the information presented here may help the reader find the basic tools to understand historical biogeography as it is practiced today, and at the same time to grasp the significant influence of software programs upon the subject.

An important web site to visit is http://evolution.genetics.washington.edu, which was created by Joe Felsenstein. This site contains abundant information about programs related to phylogenetics and those indirectly related to historical biogeography. Also, the home pages of the Tree of Life Web Project (http://tolweb.org/tree/phylogeny.html) and of the Willi Hennig Society (http://www.cladistics.org), among others, are useful sources of specific software. Sites listed were accessed January, 2003.

Panbiogeography

PHYLIP

Author: Joe Felsenstein

Address: Department of Genetics, University of Washington, Seattle, WA 98195, USA

Available at: http://evolution.genetics.washington.edu, or through http://www.cladistics.org

Applicability: The program Clique is used for track compatibility analysis

SECANT

Author: Benjamin Salisbury

Address: Genaissance Pharmaceutical Inc., Five Science Park, New Haven, CT 06511, USA

Available from the author.

Applicability: The program is used for track compatibility analysis; SECANT version 2.2 is based on an earlier program, CLINCH, by Kent Fiala

Cladistic Biogeography

COMPONENT 1.5

Author: Roderic D. M. Page

Address: Division of Environmental and Evolutionary Biology, Institute of Biomedical and Life Sciences, University of Glasgow, Glasgow, UK G12 8QQ

The program is no longer available

Applicability: Component analysis

TAS

Authors: Gareth Nelson and Pauline Ladiges

Address: School of Botany, The University of Melbourne, Victoria 3052, Australia

Available from the authors

Applicability: Three-area statement, obtaining of the area × three-area statement matrices

TASS

Authors: Gareth Nelson and Pauline Ladiges

Address: School of Botany, The University of Melbourne, Victoria 3052, Australia

Available from the authors

Applicability: Paralogy-free subtrees, obtaining of paralogy-free subtrees and the resulting area *x* component matrices

CAFCA

Author: Rino Zandee

Address: Institute of Evolutionary and Ecological Science, Van deer Klaauw Laboratory, Leiden University, the Netherlands

Available at: http://wwwbio.leidenuniv.nl/~zandee/cafca.html

Applicability: Component compatibility

Event-Based Methods

COMPONENT 2.0

Author: Roderic D. M. Page

Address: Division of Environmental and Evolutionary Biology, Institute of Biomedical and Life Sciences, University of Glasgow, Glasgow, UK G12 8QQ

Available at: http://taxonomy.zoology.gla.ac.uk/rod/cpw.html

Applicability: Reconciled trees; other applications include tree consensus, tree profile comparisons, and tree simulations

TREEMAP

Author: Roderic D. M. Page

Address: Division of Environmental and Evolutionary Biology, Institute of Biomedical and Life Sciences, University of Glasgow, Glasgow, UK G12 8QQ

Available at: http://taxonomy.zoology.gla.ac.uk/rod/treemap.html

Applicability: Reconciled trees and primary Brooks parsimony analysis

DIVA

Author: Fredrik Ronquist

Address: Evolutionary Biology Centre, Uppsala University, Norbyv 18D, SE-752 36 Uppsala, Sweden

Available by anonymous FTP from web page: http://www.ebc.uu.se/systzoo/staff/ronquist.html

Applicability: Dispersal-vicariance analysis

TREEFITTER

Author: Fredrik Ronquist

Address: Evolutionary Biology Centre, Uppsala University, Norbyv 18D, SE-752 36 Uppsala, Sweden

Available by anonymous FTP from web page: http://www.ebc.uu.se/systzoo/staff/ronquist.html

Applicability: Methods that use parsimony-based tree fitting (for example, dispersal-vicariance analysis and maximum cospeciation)

Phylogeography

GEODIS

Authors: David Posada and Alan Templeton

Address: Department of Zoology, Brigham Young University, 680 WIDB (Widtsoe Building), Provo, UT 84602–5255, USA

Available at: http://bioag.byu.edu/zoology/crandall_lab/geodis.htm

Applicability: Nested clade analysis

Programs that Apply Maximum Parsimony Algorithms

HENNIG86

Author: James S. Farris

Address: Naturhistoriska Riksmuseet, Stockholm, Sweden

Available from Arnold Kluge, Museum of Zoology, University of Michigan, Ann Arbor, MI 48109, USA, or Diane Lipscomb, Department of Biological Sciences, George Washington University, Washington, DC 20052, USA, or through links at http://www.cladistics.org

Applicability: This program is applicable to all those methods that use maximum parsimony algorithms to construct area cladograms (for example, Brooks parsimony analysis, parsimony analysis of endemicity, three-area statements, and paralogy-free subtrees)

NONA

Author: Pablo Goloboff

Address: INSUE, Fundación e Instituto Miguel Lillo, Miguel Lillo 205, 4000 San Miguel de Tucumán, Argentina

Available at: http://www.cladistics.org

Applicability: This program is applicable to all those methods that use maximum parsimony algorithms to construct area cladograms

PEE-WEE

Author: Pablo Goloboff

Address: INSUE, Fundación e Instituto Miguel Lillo, Miguel Lillo 205, 4000 San Miguel de Tucumán, Argentina

Available through the author; a demo version is available at: http://www.cladistics.org. Pee-Wee is included within the NONA software.

Applicability: This program is applicable to all those methods that use maximum parsimony algorithms to construct area cladograms

PAUP*

Author: David L. Swofford

Address: School of Computational Sciences and Information Technology, Florida State University, Tallahassee, FL 32306–4120, USA

Available at: http://paup.csit.fsu.edu

Applicability: This program is applicable to all those methods that use maximum parsimony algorithms to construct area cladograms. PAUP can produce an output of the tree/s in the format of data matrices for parsimony analysis, which means it could create primary Brooks parsimony analysis matrices.

PHYLIP: see under panbiogeography heading

Applicability: The programs PENNY and MIX, among others, are applicable to all those methods that use maximum parsimony algorithms to construct area cladograms

TNT

Authors: Pablo Goloboff, James Farris, and Kevin Nixon

Address: INSUE, Fundación e Instituto Miguel Lillo, Miguel Lillo 205 4000, San Miguel de Tucumán, Argentina

A beta version is available at: http://www.cladistics.org

Applicability: This program is applicable to all those methods that use maximum parsimony algorithms to construct area cladograms; TNT is applicable to large data sets

Phylogenetic Inference

HENNIG86: See under heading Programs that Apply Maximum Parsimony Algorithms

Applicability: Phylogenetic inference based on parsimony

NONA: See under heading Programs that Apply Maximum Parsimony Algorithms

Applicability: Phylogenetic inference based on parsimony

PEE-WEE: See under heading Programs that Apply Maximum Parsimony Algorithms

Applicability: Phylogenetic inference based on parsimony

PAUP*: See under heading Programs that Apply Maximum Parsimony Algorithms

Applicability: Phylogenetic inference based on parsimony and maximum likelihood

PHYLIP: See under heading Panbiogeography

Applicability: This is a package of programs to carry out parsimony, maximum likelihood, and distance matrix methods, among others

TNT: See under heading Programs that Apply Maximum Parsimony Algorithms

Applicability: Phylogenetic inference based on parsimony

MRBAYES

Authors: John Huelsenbeck and Fredrik Ronquist
Address: Dept. of Biology, University of Rochester, Rochester, NY 14627,

USA / Evolutionary Biology Centre, Uppsala University, Norbyv 18D, SE-752 36 Uppsala, Sweden

Available at: http://morphbank.ebc.uu.se/mrbayes/

Applicability: Bayesian inference of phylogeny using Markov Chain Monte Carlo methods

BAMBE

Authors: Donald Simon and Bret Larget

Address: Department of Mathematics and Computer Science, Duquesne University, Pittsburgh, PA 15282, USA

Available at: http://www.mathcs.duq.edu/larget/bambe.html

Applicability: Bayesian analysis of phylogenies. The main program uses a variety of Metropolis-Hastings algorithms

Interactive Tree Manipulation

WINCLADA

Author: Kevin Nixon

Address: L. H. Bailey Hortorium, Cornell University, Ithaca, NY 14853, USA

Available at: http://www.cladistics.org

Applicability: The program can read and edit data files and trees, display character state changes inferred by parsimony on diagrams of the trees, and launch runs to the programs NONA, Pee-Wee, and Hennig86. The program supports the development of Fitch and other optimizations so it could be applied to the ancestral areas technique

TREEVIEW

Author: Roderic D. M. Page

Address: Division of Environmental and Evolutionary Biology, Institute of Biomedical and Life Sciences, University of Glasgow, Glasgow, UK G12 8QQ

Available at: http://taxonomy.zoology.gla.ac.uk/rod/treeview.html

Applicability: Interactive program that can read and edit trees

MACCLADE

Authors: Wayne Maddison and David Maddison

Address: University of Arizona, Department of Entomology, Tucson, AZ 85721, USA

Available at: http://macclade.org

Applicability: The program displays character state changes on the trees. The program supports the development of Fitch and other optimizations so it could be applied to the ancestral areas technique

Molecular Clock

r8s

Author: Michael J. Sanderson

Address: University of California, Davis, Section of Evolution and Ecology, Davis, CA 95616, USA

Available at: http://ginger.ucdavis.edu/r8s/

Applicability: This is a program for estimating absolute rates of molecular evolution and divergence times on a phylogenetic tree

GLOSSARY

Algorithm: Set of rules for getting a specific output from a specific input. Any step must be accurately defined in a way that could be translated into computer language.

Allopatric: Occurring in geographically different places; ranges that are mutually exclusive.

Allopatric speciation: The differentiation of two reproductively isolated species from an initial one in geographic isolation.

Anagenesis: The process of evolution that produces entirely new levels of structural organization (grades).

Ancestor: The individual or population that gave rise to subsequent individuals or populations with different features.

Apomorphy: A derived character state.

Area of endemism: An area recognized by the congruent distribution of two or more taxa.

Barrier: Any biotic or abiotic feature that totally or partially restricts the movement (flow) of genes of individuals from one population or locality to another.

Baseline: A geographic or geological feature of an individual track such as crossing an ocean or sea basin, or a major tectonic feature (for example, a fault zone), that is interpreted as a diagnostic character (that is, spatial homology) uniting individual tracks that may otherwise have little in common (for example, no distributional congruence).

Bayesian inference of phylogeny: A method to infer phylogeny that depends only on the posterior probability of a hypothesis (tree) found by sampling the entire posterior probability distribution (tree space) and does so with minimal computation time.

Biodiversity: The variety of life at different levels of biological organization.

Biogeography: The study of the geographic distribution of organisms, both past and present.

Biological invasion: Successful establishment of introduced species.

Biota: All species inhabiting a specific region.

Camin-Sokal algorithm: A parsimony algorithm that implies strong restrictions to character state changes. This algorithm assumes irreversibility in the evolution of character states, forbidding the reversion to plesiomorphic states. Camin-Sokal algorithm requires all homoplasy to be accounted for by multiple or parallel transformations.

Center of origin: In dispersal biogeography, it is the area for which it is hypothesized that a taxon evolved and from which it dispersed to other areas.

Clade: A monophyletic lineage or group.

Cladistics: The method of reconstructing the evolutionary history (phylogeny) of a taxon by identifying the branching sequence of differentiation through analysis of shared derived character states applying the parsimony principle. Also called phylogenetic systematics.

Cladogenesis: The process of evolution that produces a series of branching events.

Cladogram: A branching diagram specifying the hierarchical relationships among taxa—a tree.

Clique: In systematics, a set of congruent characters that constitute a cladogram using character compatibility methods.

Cluster analysis: Multivariate analysis technique resulting in hierarchic groups of the study units based on resemblance, according to predefined rules.

Coevolution: Interdependent, simultaneous evolution of two unrelated taxa that exhibit a great ecological interaction (for example, a flower plant and its pollinator, a prey and its predator).

Component: A group of taxa as determined by the branching pattern of a cladogram. *See also* clade.

Congruence: Concordance within a data set or among data sets.

Consensus tree: Cladogram that shows the general concordance among two or more cladograms of the same taxa or areas.

Conservation biology: The integrated use of several social science and scientific disciplines to achieve conservation of biodiversity.

Consistency index (CI): Measure of the fit of a character relative to a given cladogram; otherwise, measure of the homoplasy of a character on a given cladogram. Numerically, it is the quotient between the minimum number of steps a character can exhibit on any cladogram (m) and the observed number of steps of this character on a given cladogram (s). The CI can be calculated for each character in particular or for all the characters as a set.

Disjunction: A discontinuous range of a monophyletic taxon in which a wide geographic distance separates at least two closely related populations.

Dispersal: The movement of organisms away from their point of origin.

Dispersion: The spatial distribution of individual organisms within a local population.

DNA sequencing: Method to determine the nucleotide sequence or order in a given fragment of DNA.

Ecological biogeography: The study of ecological factors influencing the distribution of organisms.

Endemic: Pertaining to a taxon that is restricted to a specified geographic area, such as a continent, lake, biome, or island.

Exotic species: A species introduced to one region from another geographical region through human activity (deliberately or accidentally). Introduced species.

Fitch algorithm: A parsimony algorithm that does not imply restrictions to character state changes. The Fitch algorithm is a generalization of the Wagner algorithm that allows unordered multistate character changes, meaning that it allows any transformation from one state to another in a single step. The Fitch algorithm assumes that any direction of the character state changes is equally probable.

Gene flow: The movement of alleles within a population or between populations caused by the dispersal of gametes or offspring.

Generalized track: A set of two or more individual tracks that are compatible or congruent according to a specified criterion (for example, shared baselines or compatible track geometries).

Gondwana: One of the supercontinents resulting from the breakup of Pangea. It comprised Antarctica, South America, Africa, India, Australia, and New Zealand. These landmasses were united for at least 1 billion years, but broke up during the late Mesozoic.

Haplotype: A combination of alleles of closely linked loci that are found in a single chromosome and tend to be inherited together.

Heuristic method: Any procedure of analysis that is not guaranteed to find the optimal solution to a problem.

Historical biogeography: The study of the distribution of organisms regarding its historical causes.

Homology: Two or more character states originating from the same character state in the common ancestor.

Homoplasy: Phenomena (convergence, parallelism, and reversion) that lead to similarity among character states not due to common ancestry.

Individual track: A line graph drawn on a map connecting the localities or distribution areas of a particular taxon or group of taxa.

Main massing: A locality or distribution area possessing the greatest concentration of diversity within the geographic range of a taxon.

Markov chain: A sequence of random variables in which the distribution of each random variable depends only on the value of its predecessor.

Markov chain Monte Carlo: A computational technique for the numerical evaluation of high-dimensional integrals by simulating a Markov chain on a parameter space.

Maximum likelihood: In phylogeny, optimality criterion to find a tree topology (and associated parameter estimates) that maximizes the probability of obtaining the observed data given a model of evolution.

Maximum parsimony: In phylogeny, optimality criterion to find a tree topology that minimizes the number of steps (events) required to explain the observed data set.

Metricity: Set of mathematical properties (symmetry, triangular inequality, distinguishability of nonidentical elements, and indistinguishability of identical elements) of a numerical function of pairs of points.

Metropolitan-Hastings: The main form of Markov chain Monte Carlo. The algorithm provides a probabilistic rejection rule to reject some proposed moves of an arbitrary irreducible Markov chain so that the resultant Markov chain has the desired stationary distribution.

Molecular clock: The hypothesis that molecules evolve in direct proportion to time, so that differences between homologous DNA sequences or proteins can be used to estimate the time elapsed since the two molecules (or the taxa that contain them) last shared a common ancestor.

Monophyletic: Having arisen from a common ancestral form.

Multistate character: A character that has more than two observed states. It can be considered as ordered when the sequence of character state transformation has been determined, otherwise, the character is considered as unordered. In an ordered multistate character the transformation between two adjacent states costs 1, whereas the transformation between two nonadjacent states costs the sum of the steps between their implied adjacent states (for example, transformation from state 1 to state 0 costs 1, whereas a transformation from state 0 to state 2 costs 2). In an unordered multistate character any change between two states, whether adjacent or nonadjacent states, costs the same (for example, transformation from state 1 to state 0 costs 1, also a transformation from state 0 to state 2 costs 1).

Neighbor joining: Aglomerative method of phylogenetic reconstruction, close to parsimony, that works on a distance matrix. It does not necessarily result in the shortest tree.

Null hypothesis: A statistical hypothesis stating what would be expected by chance alone, which can be tested in order to determine whether an observation could be a result of chance or is instead the result of some directing force.

Outgroup: A taxon used in phylogenetic analysis for comparative purposes, usually with respect to character polarity determination.

Panbiogeographic node: An area or locality where two or more generalized tracks overlap.

Paraphyletic: Referring to taxa that are classified chiefly on the basis of shared primitive character states.

Parsimony: A methodological principle which states that observed data should be explained in the simplest possible way (otherwise known as Ockham's razor).

Pattern: Nonrandom, repetitive organization.

Phylogeny: The evolutionary relationships between an ancestor and all of its known descendants.

Plate tectonics: The study of the origin, movement, and involvement in the evolution of the Earth's crust.

Plesiomorphy: A primitive or ancestral character state.

Polymerase chain reaction (PCR): A process for amplifying a target DNA sequence, in which a series of thermal cycles each result in denaturation of a double-stranded target, annealing of oligonucleotide primers to the resulting single strands, and primer extension catalyzed by a thermostable DNA polymerase.

Posterior distribution: The conditional probability distribution of one or more parameters, after observation of data.

Restriction endonucleases: Enzymes isolated from bacteria that cut DNA at a constant position within a specific recognition sequence (restriction site), typically 4–6 bp long.

Retention index (RI): Measure of the fit of a character relative to a given cladogram that takes into account the greatest number of steps a character can exhibit in any cladogram (g). Numerically, the RI is the ratio between ($g-s$) and ($g-m$), where m is the minimum number of steps a character can exhibit on any cladogram and s is the observed number of steps of this character on a given cladogram. The RI can be calculated for each character or for all the characters as a set.

Sister taxa (sister areas): The two taxa (or areas) that are most closely (and therefore more recently) related.

Speciation: The process in which two or more contemporaneous species evolve from a single ancestral population.

Stochastic: Random, expected (statistically) by chance alone.

Sympatric: Living in the same local community, close enough to interact. In the more general sense, having broadly overlapping geographic distributions.

Sympatric speciation: The differentiation of two reproductively isolated species from one initial population within the same local area; hence, speciation that occurs under conditions in which much gene flow potentially could or actually does occur.

Synapomorphy: A derived character state shared among taxa.

Taxon (pl. taxa): A group of organisms considered as a unit of any rank in a classificatory system. Examples are the family Apidae, the order Primates, and the genus *Nassauvia.*

Vagility: The ability to move actively from one place to another.

Vicariance biogeography: A branch of historical biogeography that attempts to reconstruct the historical events that led to the observed distributional patterns based largely on the assumption that these patterns resulted from the splitting (vicariance) of areas and not long dispersal.

Vicariance event: Split of a taxon or biota in two or more geographic subsets due to the appearance of a natural barrier (for example, glaciation, continental drift).

Vicariants: Two disjunct taxa that are closely related to each other and that are assumed to have originated when the initial range of their ancestor was split by some historical event.

WORKS CITED

Aares, E., M. Nurminiemi, and C. Brochmann. 2000. Incongruent phylogeographies in spite of similar morphology, ecology, and distribution: *Phippsia algida* and *P. concinna* (Poaceae) in the North Atlantic region. *Plant Syst. Evol.* 220:241–261.

Aguilar-Aguilar, R. and R. Contreras-Medina. 2001. La distribución de los mamíferos marinos en México: Un enfoque panbiogeográfico. In *Introducción a la Biogeografía en América Latina: Teorías, Conceptos, Métodos y Aplicaciones,* J. Llorente-Bousquets and J. J. Morrone (eds.). México: Las Prensas de Ciencias, pp. 213–220.

Alroy, J. 1995. Continuous track analysis: A new phylogenetic and biogeographic method. *Syst. Biol.* 44:152–187.

Andersson, L. 1996. An ontological dilemma: Epistemology and methodology of historical biogeography. *J. Biogeog.* 23:269–277.

Artigas, J. N. 1975. Introducción al estudio por computación de las áreas zoogeográficas de Chile continental basado en la distribución de 903 especies de animales terrestres. *Gayana,* Misc. 4:1–25.

Avise, J. C. 1992. Molecular population structure and the biogeographic history of a regional fauna: A case history with lessons for conservation biology. *Oikos* 63:62–76.

——— 1994. *Molecular Markers, Natural History and Evolution.* New York: Chapman & Hall.

——— 2000. *Phylogeography: The History and Formation of Species.* Cambridge, Mass.: Harvard University Press.

Avise, J. C., J. Arnold, R. M. Ball, E. Bermingham, T. Lamb, J. E. Neigel, C. A. Reeb, and N. C. Saunders. 1987. Intraspecific phylogeography: The mitochondrial DNA bridge between population genetics and systematics. *Annu. Rev. Ecol. Syst.* 18:489–522.

Axelius, B. 1991. Areas of distribution and areas of endemism. *Cladistics* 7:191–199.

Baldwin, B. G. and R. H. Robichaux. 1995. Historical biogeography and ecology of the Hawaiian silversword alliance (Asteraceae). In *Hawaiian Biogeography: Evolution on a Hot Spot Archipelago,* W. L. Wagner and V. A. Funk (eds.). Washington and London: Smithsonian Institution Press, pp. 259–287.

Ball, I. R. 1990. The framing of biogeographic hypotheses. In *Evolutionary Biogeography of he Marine Algae of the North Atlantic,* D. J. Garbary and G. R. South (eds.). Berlin Heidelberg: Springer-Verlag, pp. 1–7.

Bates, J. M., S. J. Hackett, and J. Cracraft. 1998. Area-relationships in the Neotropical lowlands: An hypothesis based on raw distributions of Passerine birds. *J. Biogeog.* 25:783–793.

Beyra, M. A. and M. Lavin. 1999. Monograph of *Pictetia* (Leguminosae-Papilionoideae) and review of the Aeschynomeneae. *Syst. Bot. Monographs* 56:1–93.

Bisconti, M., W. Landini, G. Bianucci, G. Cantalamessa, G. Carnevale, L. Ragaini, and G. Valleri. 2001. Biogeographic relationships of the Galapagos terrestrial biota: Parsimony analyses of endemicity based on reptiles, land birds and *Scalesia* land plants. *J. Biogeog.* 28:495–510.

Blake, R. D., T. H. Samuel, and J. Nicholson-Tuell. 1992. The influence of nearest neighbors on rate and pattern of spontaneous point mutations. *J. Mol. Evol.* 34:189–200.

Bremer, K. 1992. Ancestral areas: A cladistic reinterpretation of the center of origin concept. *Syst. Biol.* 41:436–445.

——— 1993. Intercontinental relationships of African and South American Asteraceae: A cladistic biogeographic analysis. In *Biological Relationships between Africa and South America,* P. Goldblatt (ed.). New Haven: Yale University Press, pp. 105–135.

——— 1995. Ancestral areas: Optimization and probability. *Syst. Biol.* 44:255–259.

Brooks, D. R. 1985. Historical ecology: A new approach to studying the evolution of ecological associations. *Ann. Missouri Bot. Gard.* 72:660–680.

——— 1988. Scaling effects in historical biogeography: A new view of space, time, and form. *Syst. Zool.* 32:237–244.

——— 1990. Parsimony analysis in historical biogeography and coevolution: Methodological and theoretical update. *Syst. Zool.* 39:14–30.

Brooks, D. R. and D. A. McLennan. 1991. *Phylogeny, Ecology, and Behavior. A Research Program in Comparative Biology.* Chicago: University of Chicago Press.

——— 2001. A comparison of a discovery-based and an event-based method of historical biogeography. *J. Biogeog.* 28:757–767.

Brooks, D. R., M. P. G. van Veller, and D. A. McLennan. 2001. How to do BPA, really. *J. Biogeog.* 28:345–358.

Brown, G. K., F. Udovicic, and P. Y. Ladiges. 2001. Molecular phylogeny and biogeography of *Melaleuca, Callistemon* and related genera (Myrtaceae). *Aust. Syst. Bot.* 14:565–585.

Brown, J. H. 1989. Patterns, modes and extents of invasions by vertebrates. In *Biological Invasions: A Global Perspective,* J. A. Drake, H. A. Mooney, F. di Castri, R. H. Groves, F. J. Kruger, M. Rejmánek, and M. Williamson (eds.). New York: Wiley & Sons, pp. 85–109.

Brown, J. H. and M. B. Lomolino. 1998. *Biogeography,* 2nd Ed. Sunderland, Mass.: Sinauer.

Brown, J. M., J. H. Leebens-Mack, J. N. Thompson, O. Pellmyr, and R. G. Harrison. 1997. Phylogeography and host association in a pollinating seed parasite *Greya politella* (Lepidoptera: Prodoxidae). *Mol. Ecol.* 6:215–224.

Brundin, L. 1965. On the real nature of Transantarctic relationships. *Evolution* 19:496–505.

——— 1966. Transantarctic relationships and their significance. *Kungl. Svens. Vetenskapakad. Handl.* 11:1–472.

——— 1972. Phylogenetics and biogeography. *Syst. Zool.* 21:69–79.

——— 1981. Croizat's panbiogeography versus phylogenetic biogeography. In *Vicariance Biogeography: A Critique,* G. Nelson and D. E. Rosen (eds.). New York: Columbia University Press, pp. 94–158.

Cabrera A. L. and A. Willink. 1973. *Biogeografía de América Latina.* Washington, D. C.: O.E.A. Serie de Biología, Monografía 13.

Caccone, A., M. C. Milinkovitch, V. Sbordoni, and J. R. Powell. 1994. Molecular biogeography: Using the Corcica-Sardinia microplate disjunction to calibrate mitochondrial rDNA evolutionary rates in mountain newts *(Euproctus). J. Evol. Biol.* 7:227–245.

——— 1997. Mitochondrial DNA rates and biogeography in European newts (genus *Euproctus*). *Syst. Biol.* 46:126–144.

Cadle, J. E. and H. W. Greene. 1994. Phylogenetic patterns, biogeography, and the ecological structure of neotropical snake assemblages. In *Species Diversity in Ecological Communities: Historical and Geographical Perspectives,* R. E. Ricklefs and D. Schulter (eds.). Chicago: The University of Chicago Press, pp. 281–293.

Cain, S. A. 1944. *Foundations of Plant Geography.* New York: Harper & Brothers.

Candolle, A. P. de. 1820. Géographie botanique. *Dict. Sci. Nat.* 18:359–422.

——— 1838. *Statistique de la Familie des Composées.* Paris and Strasburgo: Treutel & Würtz.

Cavalli-Sforza, L. L. and A. W. F. Edwards. 1967. Phylogenetic analysis: Models and estimation procedures. *Evolution* 32:550–570.

Cavieres, L. A., M. T. K. Arroyo, P. Posadas, C. Marticorena, O. Matthei, R. Rodríguez, F. A. Squeo, and G. Arancio. 2002. Identification of priority areas for conservation in an arid zone: Application of parsimony analysis of endemicity in the vascular flora of the Antofagasta region, northern Chile. *Biodiversity and Conservation* 11:1301–1311.

Chanderbali, A. S., H. van der Werff, and S. S. Renner. 2001. Phylogeny and historical biogeography of Lauraceae: Evidence from the chloroplast and nuclear genomes. *Ann. Missouri Bot. Gard.* 88:104–134.

Chapco, W., R. A. Kelln, and D. A. McFayden. 1992. Intraspecific mitochondrial DNA variation in the migratory grasshopper, *Melanoplus sanguinipes. Heredity* 69:547–557.

Charleston, M. A. 1998. Jungles: A new solution to the host/parasite phylogeny reconciliation problem. *Math. Biosci.* 149:191–223.

Climo, F. M. 1989. The panbiogeography of New Zealand as illuminated by the genus *Fectola* Iredale, 1915 and subfamily Rotadiscinae Pilsbry, 1927 (Mollusca: Pulmonata: Punctoidea: Charopidae). *New Zealand J. Zool.* 16:587–649.

Condie, K. C. 1997. *Plate Tectonics and Crustal Evolution,* 4th ed. Woburn, Mass.: Butterworth & Heinemann.

Connor, E. F. 1988. Fossils, phenetics and phylogenetics: Inferring the historical dynamics of biogeographic distributions. In *Zoogeography of Caribbean Insects,* J. K. Liebherr (ed.). Ithaca, N.Y.: Cornell University Press, pp. 254–269.

Conran, J. G. 1995. Family distributions in the Liliiflorae and their biogeographical implications. *J. Biogeog.* 22:1023–1034.

Contreras-Medina, R. and H. Eliosa. 2001. Una visión panbiogeográfica preliminar de México. In *Introducción a la Biogeografía en América Latina: Teorías, Conceptos, Métodos y Aplicaciones,* J. Llorente-Bousquets and J. J. Morrone (eds.). México: Las Prensas de Ciencias, pp. 197–212.

Couper, R. A. 1960. Southern hemisphere Mesozoic and Terciary Podocarpaceae and Fagaceae and their palaeogeographic significance. *Proc. R. Soc. B.* 152:491–500.

Cox, C. B. and P. D. Moore. 1993. *Biogeography: An Ecological and Evolutionary Approach,* 5th ed. Oxford: Blackwell.

Cracraft, J. 1991. Patterns of diversification within continental biotas: Hierarchical congruence among the areas of endemism of Australian vertebrates. *Aust. Syst. Bot.* 4:211–227.

Craw, R. C. 1979. Generalized tracks and dispersal in biogeography: A response to R. M. McDowall. *Syst. Zool.* 28:99–107.

——— 1982. Phylogenetics, areas, geology, and the biogeography of Croizat: A radical view. *Syst. Zool.* 31:304–316.

——— 1983. Panbiogeography and vicariance cladistics: Are they truly different? *Syst. Zool.* 32:431–438.

——— 1984a. Biogeography and biogeographic principles. *New Zealand Entomol.* 8:49–52.

——— 1984b. Leon Croizat's biogeographic work: A personal appreciation. *Tuatara* 27:8–13.

——— 1985. Classic problems of southern hemisphere biogeography re-examined: Panbiogeographic analysis of the New Zealand frog *Leiopelma*, the ratite birds and *Nothofagus*. *Z. Zool. Syst. Evolutionsforsch.* 23:1–10.

——— 1988a. Continuing the synthesis between panbiogeography, phylogenetic systematics and geology as illustrated by empirical studies on the biogeography of New Zealand and the Chatham Islands. *Syst. Zool.* 37:291–310.

——— 1988b. Panbiogeography: Method and synthesis in biogeography. In *Analytical Biogeography: An Integrated Approach to the Study of Animal and Plant Distributions,* A. A. Myers and P. S. Giller (eds.). New York: Chapman & Hall, pp. 405–435.

——— 1989. Quantitative panbiogeography: Introduction to methods. *New Zealand J. Zool.* 16:485–494.

——— 1990. New Zealand biogeography: A panbiogeographic approach. *New Zealand J. Zool.* 16:527–547.

Craw, R. C., J. R. Grehan, and M. J. Heads. 1999. *Panbiogeography: Tracking the History of Life.* New York: Oxford University Press.

Craw, R. C. and R. D. M. Page. 1988. Panbiogeography: Method and metaphor in the new biogeography. In *Evolutionary Processes and Metaphors,* M. W. Ho and S. W. Fox (eds.). Chichester, U.K.: John Wiley & Sons, pp. 163–189.

Craw, R. C. and P. Weston. 1984. Panbiogeography: A progressive research program? *Syst. Zool.* 33:1–33.

Crisci, J. V. 1981. La especie: Realidad y conceptos. Symposia, *VI Jornadas Argentinas de Zoología,* La Plata, Argentina, pp. 21–32.

——— 1982. Parsimony in evolutionary theory: Law or methodological prescription? *J. Theor. Biol.* Special issue on unsolved problems 97:35–41.

——— 1992. Reflexiones en torno a la Biología Comparada. *Anal. Acad. Nac. Cs. Ex. Fís. Nat.* 44:159–164.

——— 1998a. Forma, espacio, tiempo: Los métodos modernos de reconstrucción filogenética. *Monogr. Syst. Bot. Missouri Bot. Gard.* Proceedings of the VI Congreso Latinoamericano de Botánica 68:443–444.

——— 1998b. El cladismo y la biogeografía histórica. *Monogr. Syst. Bot. Missouri Bot. Gard.* Proceedings of the VI Congreso Latinoamericano de Botánica 68:459–463.

——— 1998c. La sistemática de nuestro tiempo: Hechos, problemas y orientaciones. *Bol. Soc. Bot. México* 63:21–32.

——— 2001. The voice of historical biogeography. *J. Biogeog.* 28:157–168.

Crisci, J. V., M. M. Cigliano, J. J. Morrone, and S. Roig-Juñent. 1991a. Historical biogeography of southern South America. *Syst. Zool.* 40:152–171.

——— 1991b. A comparative review of cladistic biogeography approaches to historical biogeography of southern South America. *Aust. Syst. Bot.* 4:117–126.

Crisci, J. V., S. E. Freire, G. Sancho, and L. Katinas. 2001. Historical biogeography of the Asteraceae from Tandilia and Ventania mountain ranges (Buenos Aires, Argentina). *Caldasia* 23:21–41.

Crisci, J. V. and L. Katinas. 1997. La filogenia frente a la justicia. *Ciencia Hoy* 6:34–40.

Crisci, J. V., L. Katinas, and P. Posadas. 2000. *Introducción a la Teoría y Práctica de la Biogeografía Histórica.* Buenos Aires, Argentina: Sociedad Argentina de Botánica.

Crisci, J. V., A. A. Lanteri, and E. Ortiz Jaureguizar. 1994. Programas de computación en sistemática y biogeografía histórica: Revisión crítica y criterios para su selección. In *Taxonomía Biológica,* J. Llorente-Bousquets and I. Luna (eds.). México: UNAM, Fondo de Cultura Económica, pp. 207–225.

Crisci, J. V. and J. J. Morrone. 1990. En busca del paraíso perdido: La biogeografía histórica. *Ciencia Hoy* 1:26–34.

——— 1992a. A comparison of biogeographic models: A response to Bastow Wilson. *Glob. Ecol. Biodiv. Lett.* 2:174–176.

——— 1992b. Panbiogeografía y biogeografía cladística: Paradigmas actuales de la biogeografía histórica. *Ciencias* (México), n. esp. 6:87–97.

Crisci, J. V., P. Posadas, L. Katinas, and D. R. Miranda-Esquivel. 1999. Estrategias evolutivas para la conservación de la biodiversidad en América del Sur austral. In *Biodiversidad y Uso de la Tierra: Conceptos y Ejemplos de Latinoamérica,* S. D. Matteucci, O. T. Solbrig, J. Morello, and G. Halffter (eds.). Buenos Aires, Argentina: EUDEBA-UNESCO, pp. 175–198.

Crisci, J. V., P. Posadas, and J. J. Morrone. 1996. La biodiversidad en los umbrales del siglo XXI. *Ciencia Hoy* 6:34–40.

Crisci, J. V. and T. F. Stuessy. 1980. Determining primitive character states for phylogenetic reconstruction. *Syst. Bot.* 5:112–135.

Crisp, M. D. 2001. Historical biogeography and patterns of diversity in plants, algae and fungi: Introduction. *J. Biogeog.* 28:153–155.

Crisp, M. D., H. P. Linder, and P. H. Weston. 1995. Cladistic biogeography of plants in Australia and New Guinea: Congruent pattern reveals two endemic tropical tracks. *Syst. Biol.* 44:457–473.

Croizat, L. 1952. *Manual of Phytogeography.* Junk, The Hague.

——— 1958. *Panbiogeography,* Vols. I, IIa, and IIb. Caracas, Venezuela: Published by the author.

——— 1964. *Space, Time, Form: The Biological Synthesis.* Caracas, Venezuela: Published by the author.

——— 1981. Biogeography: Past, present, and future. In *Vicariance Biogeography: A Critique,* G. Nelson and D. E. Rosen (eds.). New York: Columbia University Press, pp. 501–523.

——— 1982. Vicariance/vicariism, panbiogeography, "vicariance biogeography"; etc.: A clarification. *Syst. Zool.* 31:291–304.

Croizat, L., G. Nelson, and D. E. Rosen. 1974. Centers of origin and related concepts. *Syst. Zool.* 23:265–287.

Croizer, R. H. 1992. Genetic diversity and the agony of choice. *Biol. Conserv.* 61:11–15.

——— 1997. Preserving the information content of species: Genetic diversity, phylogeny, and conservation worth. *Ann. Rev. Ecol. Syst.* 28:243–268.

Cruzan, M. B. 1999. Intraspecific diversification: Plants as model systems for phylogeographic studies. Abstracts, *XVI International Botanical Congress,* St. Louis, Missouri, p. 257.

Darlington, P. J. Jr. 1957. *Zoogeography: The Geographical Distribution of Animals.* New York: Wiley.

——— 1965. *Biogeography of the Southern End of the World: Distribution and History of Far-Southern Life and Land, with an assessment of Continental Drift.* Cambridge, Mass.: Harvard University Press.

——— 1970. A practical criticism of Hennig-Brundin "Phylogenetic systematics" and Antarctic biogeography. *Syst. Zool.* 19:1–18.

Darwin, C. 1859. *On the Origin of Species by Means of Natural Selection or the Preservation of Favoured Races on the Struggle for Life.* London: John Murray.

Davis, C. C., C. D. Bell, S. Mathews, and M. J. Donoghue. 2002. Laurasian migration explains Gondwanan disjunctions: Evidence from Malpighiaceae. *Proc. Nat. Acad. Sci. USA* 95:9402–9406.

Dawson, M. N. 2001. Phylogeography in coastal marine animals: A solution from California? *J. Biogeog.* 28:723–736.

De Meyer, M. 1996. Cladistic and biogeographic analysis of Hawaiian Pipunculidae (Diptera) revisited. *Cladistics* 12:291–303.

Desalle, R. 1995. Molecular approaches to biogeographic analysis of Hawaiian Drosophilidae. In *Hawaiian Biogeography: Evolution on a Hot Spot Archipelago,* W. L. Wagner and V. A. Funk (eds.). Washington and London: Smithsonian Institution Press, pp. 72–89.

Dettman, M. E. 1989. Antarctica: Cretaceous cradle of austral temperate rainforests? In *Origins and Evolution of the Antarctic Biota,* J. A. Crame (ed.). Geological Society Special Publications 47:89–105.

di Castri, F. 1989. History of biological invasions with special emphasis in the Old World. In *Biological Invasions: A Global Perspective,* J. A. Drake, H. A. Mooney, F. di Castri, R. H. Groves, F. J. Kruger, M. Rejmánek, and M. Williamson (eds.). New York: Wiley & Sons, pp. 1–30.

Domínguez, E. 1999. Systematics, cladistics and biogeography of the American genus *Farrodes* (Ephemeroptera: Leptophlebiidae: Atalophlebiinae). *Zool. J. Linn. Soc.* 126:155–189.

Donoghue, M. J., C. D. Bell, and J. Li. 2001. Phylogenetic patterns in Northern Hemisphere plant geography. *Int. J. Plant. Sci.* 162:41–52.

Ebach, M. C. 1999. Paralogy and the centre of origin concept. *Cladistics* 15:387–391.

Ebach, M. C. and G. D. Edgecombe. 2001. Cladistic biogeography: Component-based methods and paleontological application. In *Fossils, Phylogeny and Form,* J. M. Adrain, G. D. Edgecombe, and B. S. Lieberman (eds.). New York: Plenum Publishers, pp. 235–289.

Ebenhard, T. 1988. Introduced bird and mammals and their ecological effects. *Swedish Wildlife Res.* 13:1–107.

Edwards, A. W. F. 1996. The origin and early development of the method of minimum evolution for the reconstruction of phylogenetic trees. *Syst. Biol.* 45:79–91.

Edwards, A. W. F. and L. L. Cavalli-Sforza. 1964. Reconstruction of evolutionary trees. In *Phenetic and Phylogenetic Classification,* V. H. Heywood and J. McNeill (eds.). London: Systematics Association, pp. 67–76.

Ellsworth, D. L., R. L. Honeycutt, N. J. Silvy, J. W. Bickham, and J. W. Klimstra. 1994. Historical biogeography and contemporary patterns of mitochondrial DNA variation in white-tailed deer from southestern United States. *Evolution* 48:122–136.

Emerson, B. C., G. P. Walis, and B. H. Patricks. 1997. Biogeographic area relationships in southern New Zealand: A cladistic analysis of Lepidoptera distributions. *J. Biogeog.* 24:89–99.

Enghoff, H. 1996. Widespread taxa, sympatry, dispersal, and an algorithm for resolved area cladograms. *Cladistics* 12:349–364.

Engler, A. 1882. *Versuch einer Entwicklungsgeschiete der Pflanzenwelt.* Leipzig: Engelmann.

Ewel, J. J., D. J. O'Dowd, J. Bergelson, C. C. Daehler, C. M. D'Antonio, L. D. Gómez, D. R. Gordon, R. J. Hobbs, A. Holt, K. R. Hopper, C. E. Hughes, M. LaHart, R. R. B. Leakey, W. G. Lee, L. L. Loope, D. H. Lorence, S. M. Louda, A. E. Lugo, P. B. Mc Evoy, D. M. Richardson, and P. M. Vitousek. 1999. Deliberate introductions of species: Research needs. *Bioscience* 49:619–630.

Ezcurra, C., A. Ruggiero, and J. V. Crisci. 1997. Phylogeny of *Chuquiraga* sect. *Acanthophyllae* (Asteraceae-Barnadesioideae), and the evolution of its leaf morphology in relation to climate. *Syst. Bot.* 22:151–163.

Faith, D. P. 1992a. Conservation evaluation and phylogenetic diversity. *Biol. Conserv.* 61:1–10.

——— 1992b. Systematics and conservation: On predicting the feature diversity of subsets of taxa. *Cladistics* 8:361–373.

——— 1993. Biodiversity and systematics: The use and misuse of divergence information in assessing taxonomic diversity. *Pacific Conserv. Biol.* 1:53–57.

——— 1994a. Genetic diversity and taxonomic priorities for conservation. *Biol. Conserv.* 68:69–74.

——— 1994b. Phylogenetic pattern and the quantification of organismal diversity. *Philos. Trans. R. Soc. London Ser. B* 345:45–58.

——— 1994c. Phylogenetic diversity: A general framework for the prediction of feature diversity. In *Systematics and Conservation Evaluation,* P. L. Forey, C. J. Humphries, and R. I. Vane-Wright (eds.). Oxford: Clarendon Press, pp. 251–268.

Farris, J. S. 1988. *Hennig86 reference, version 1.5.* Port Jefferson, New York: Published by the author.

Felsenstein, J. 1981. Evolutionary trees from DNA sequences: A maximum likelihood approach. *J. Mol. Evol.* 17:368–376.

——— 1985. Confidence limits on the phylogenies: An approach using the boostrap. *Evolution* 39:783–791.

——— 1993. *PHYLIP (Phylogeny Inference Package), version 3.5c.* Seattle: Department of Genetics, University of Washington.

——— 2001. The troubled growth of statistical phylogenetics. *Syst. Biol.* 50:465–467.

Fiala, K. 1984. *CLINCH (CLadistic INference by Compatibility of Hypothesized characters), version 6.2. User instruction.* Stony Brook: SAS Institute.

Fitch, W. M. 1971. Toward defining the course of evolution: Minimum change for a specific tree topology. *Syst. Zool.* 20:406–416.

Florin, R. 1940. The Tertiary fossil conifers of south Chile and their phytogeographical significance. *Kungl. Svens. Vetenskapakad. Handl.* 14:1–107.

Forey, P. L., C. J. Humphries, I. J. Kitching, R. W. Scotland, D. J. Siebert, and D. M. Williams. 1992. *Cladistics: A Practical Course in Systematics.* Oxford: Clarendon Press.

Fos, M., M. A. Domínguez, A. Latorre, and A. Moya. 1990. Mitochondrial DNA evolution in experimental populations of *Drosophila subobscura. Proc Nat. Acad. Sciences* 87: 4198–4201.

Francisco-Ortega, J., D. J. Crawford, A. Santos-Guerra, and R. K. Jansen. 1997. Origin and evolution of *Argyranthemum* (Asteraceae: Anthemidae) in Macaronesia. *In Molecular Evolution and Adaptive Radiation,* T. J. Givinish and K. J. Sytsma (eds.). Cambridge: Cambridge University Press, pp. 407–431.

Franco, P. 2001. Estudios panbiogeográficos en Colombia. In *Introducción a la Biogeografía en América Latina: Teorías, Conceptos, Métodos y Aplicaciones,* J. Llorente-Bousquets and J. J. Morrone (eds.). México: Las Prensas de Ciencias, pp. 221–224.

Frenguelli, J. 1950. Rasgos generales de la morfología y la geología de la provincia de Buenos Aires. La Plata, Argentina: L.E.M.I.T. Serie 2, n. 33, pp. 72.

Fritsch, P. 1999. Phylogeny of *Styrax* based on morphological characters, with implications for biogeography and infrageneric classification. *Syst. Bot.* 24:356–378.

——— 2001. Phylogeny and biogeography of the flowering plant genus *Styrax* (Styracaceae) based on chloroplast DNA restriction sites and DNA sequences of the internal transcribed spacer region. *Mol. Phylog. Evol.* 19:387–408.

Fujii, N., N. Tomaru, K. Okuyama, T. Koike, T. Mikami, and K. Ukeda. 2002. Chloroplast DNA phylogeography of *Fagus crenata* (Fagaceae) in Japan. *Plant. Syst. Evol.* 232:21–33.

García-Barros, E., P. Gurrea, M. J. Luciáñez, J. M. Cano, M. L. Munguira, J. C. Moreno, H. Sainz, M. J. Sanz, and J. C. Simón. 2002. Parsimony analysis of endemicity and its application to animal and plant geographical distribution in the Ibero-Balearic region (western Mediterranean). *J. Biogeog.* 29:109–124.

Gatrell, A. 1983. *Distance and Space: A Geographical Perspective.* Oxford: Clarendon Press.

Gillespie, J. H. 1991. *The Causes of Molecular Evolution.* New York: Oxford University Press.

Goldman, N. 1993. Statistical tests of models of DNA substitution. *J. Molec. Evol.* 36:182–198.

Goloboff, P. A. 1996. *NONA, versión 1.5.1.* Tucumán, Argentina: Published by the author.

Grehan, J. R. 1988a. Panbiogeography: Evolution in space and time. *Riv. Biol., Biol. Forum* 81:469–498.

——— 1988b. Biogeographic homology: Ratites and the southern beeches. *Riv. Biol., Biol. Forum* 81:577–587.

——— 1991. Panbiogeography 1981–91: Development of an earth/life synthesis. *Progr. Phys. Geogr.* 15:331–363.

——— 1993. Conservation biogeography and the biodiversity crisis: A global problem in space/time. *Biodiversity Lett.* 1:134–140.

——— 2000. Atlas of global biodiversity: Mapping the spatial structure of life. *Biodiversity* 1:21–24.

——— 2001a. Panbiogeography from tracks to ocean basins: Evolving perspectives. *J. Biogeog.* 28:413–429.

——— 2001b. Biogeography and evolution of the Galapagos: Integration of the biological and geological evidence. *Biol. J. Linnean Soc.* 74: 267–287.

Hadju, E. 1995. *Macroevolutionary Patterns within the Demosponge Order Poecilosclerida: Phylogeny of the Marine Cosmopolitan Genus Mycale, and an Integrated Approach to Biogeography of the Seas.* Centrale Drukkerij: University of Amsterdam.

——— 1998. Toward a panbiogeography of the seas: Sponge phylogenies and general tracks. In *Sponge Sciences: Multidisciplinary Perspectives,* Y. Watanabe and N. Fusetani (eds.). Tokyo: Springer-Verlag, pp. 95–108.

Haeckel, E. 1866. *Generelle Morphologie der Organismen: Allgemeine Grundzüge der organischen Formen-Wissenschaft, mechanisch begründet durch die von Charles Darwin reformirte Descendenz-Theorie,* 2 vols. Berlin: Georg Reimer.

Hahn, W. J. and K. J. Sytsma. 1999. Molecular systematics and biogeography of the southeast Asian genus *Caryota* (Palmae). *Syst. Bot.* 24:558–580.

Hall, B. G. 2001. *Phylogenetic Trees Made Easy: A How-to Manual for Molecular Biologists.* Sunderland, Mass.: Sinauer.

Hanks, S. L. and D. E. Fairbrothers. 1976. Palynotaxonomic investigation of *Fagus* L. and *Nothofagus* Bl.: Light microscopy, scanning electron microscopy and computer analyses. In *Botanical Systematics: An Occasional Series of Monographs,* V. H. Heywood (ed.). London and New York: Academic Press, Vol. 1, pp. 1–141.

Haraway, D. J. 1976. *Crystal, Fabrics, and Fields: Metaphors of Organicism in Twentieth-Century Developmental Biology.* New Haven and London: Yale University Press.

Harold, A. S. and R. D. Mooi. 1994. Areas of endemism: Definition and recognition criteria. *Syst. Biol.* 43:261–266.

Harvey, P. H. and S. Nee. 1996. What this book is about. In *New Uses for New Phylogenies,* P. H. Harvey, A. J. Leigh Brown, J. Maynard Smith, and S. Nee (eds.). New York: Oxford University Press, pp. 1–11.

Hausdorf, B. 1998. Weighted ancestral area analysis and a solution of the redundant distribution problem. *Syst. Biol.* 47:445–456.

——— 2000. Biogeography of the Limacoidea sensu lato (Gastropoda: Stylommatophora): Vicariance events and long-distance dispersal. *J. Biogeog.* 27:379–390.

——— 2002. Units in biogeography. *Syst. Biol.* 51:648–652.

Haydon, D. T., B. I. Crother, and E. R. Pianka. 1994. New directions in biogeography? *Trends Ecol. Evol.* 10:403–406.

Haydon, D. T., R. R. Radtkey, and E. R. Pianka. 1994. Experimental biogeography: Interactions between stochastic, and ecological processes in a model archipelago. In *Species Diversity in Ecological Communities: Historical and Geographical Perspectives,* R. E. Ricklefs and D. Schulter (eds.). Chicago: University of Chicago Press, pp. 117–130.

Hayes, J. P. and R. G. Harrison. 1992. Variation in mitochondrial DNA and the biogeographic history of woodrats (*Neotoma*) of the eastern United States. *Syst. Biol.* 41:331–344.

Heads, M. J. 1986. A panbiogeographic analysis of Auckland islands archipelago. In *The Lepidoptera, Bryophytes, and Panbiogeography of Auckland Islands,* R. D. Archibald (ed.). Dunedin, New Zealand: New Zealand Entomological Society, pp. 30–44.

——— 1990. Integrated earth and life sciences in New Zealand natural history: The parallel arcs model. *New Zealand J. Zool.* 16:549–585.

——— 1999. Vicariance biogeography and terrane tectonics in the South Pacific: An analysis of the genus *Abrotanella* (Compositae), with a new species from New Zealand. *Biol. J. Linnean Soc.* 67:391–492.

——— 2001. Birds of paradise, biogeography and ecology in New Guinea: A review. *J. Biogeog.* 28:893–926.

Heard, S. B. and A. Ø. Mooers. 2000. Phylogenetically patterned speciation rates and extinction risks change the loss of evolutionary history during extinctions. *Proc. R. Soc. London Ser. B* 267:613–620.

Heisenberg, W. 1958. *The Physicist´s Conception of Nature.* New York: Harcourt Brace.

Helfgott, D. M., J. Francisco-Ortega, A. Santos-Guerra, R. K. Jansen, and B. B. Simpson. 2000. Biogeography and breeding system evolution of the woody *Bencomia* alliance (Rosaceae) in Macaronesia based on ITS sequence data. *Syst. Bot.* 25:82–97.

Henderson, I. M. 1990. Quantitative biogeography: An investigation into concepts and methods. *New Zealand J. Zool.* 16:495–510.

——— 1991. Biogeography without areas? *Aust. Syst. Bot.* 4:59–71.

Hennig, W. 1950. *Grundzüge Einer Theorie der Phylogenetischen Systematics.* Urbana: University of Illinois Press.

——— 1966. *Phylogenetic Systematics.* Urbana: University of Illinois Press.

Hesse, M. 1966. *Models and Analogies in Science.* South Bemd, Indiana: University of Notre Dame Press.

Heywood, V. H. 1993. *Flowering Plants of the World.* Updated edition. New York: Oxford University Press.

Hibbett, D. S. 2001. Shiitake mushrooms and molecular clock: Historical biogeography of *Lentinula. J. Biogeog.* 28:231–241.

Hill, R. S. 1992. *Nothofagus:* Evolution from a southern perspective. *Trends Ecol. Evol.* 7:190–194.

——— 1996. The riddle of unique southern hemisphere *Nothofagus* on southwest Pacific islands: Its challenge to biogeographers. In *The Origin and Evolution of Pacific Island Biotas, New Guinea to Eastern Polynesia: Patterns and Processes,* A.

Keast and S. E. Miller (eds.). The Netherlands: SPB Academic Publishing, pp. 247–260.

——— 2001. Biogeography, evolution and palaeecology of *Nothofagus* (Nothofagaceae): The contribution of the fossil record. *Aust. J. Bot.* 49:321–332.

Hill, R. S. and M. E. Dettman. 1996. Origin and diversification of the genus *Nothofagus.* In *The Ecology and Biogeography of Nothofagus Forests,* T. T. Veblen, R. S. Hill, and J. Read (eds.). New Haven: Yale University Press, pp. 11–24.

Hill, R. S. and G. J. Jordan. 1993. The evolutionary history of *Nothofagus* (Nothofagaceae). *Aust. Syst. Bot.* 6:111–126.

Hill, R. S. and P. H. Weston. 2001. Southern (Austral) ecosystems. In *Encyclopedia of Biodiversity,* S. A. Levin (ed.). New York: Academic Press, vol. 5, pp. 361–370.

Hillis, D., B. K. Mable, and C. Moritz. 1996b. Applications of molecular systematics: The state of the field on a look to the future. In *Molecular Systematics,* 2nd Ed., D. M. Hillis, C. Moritz, and B. K. Mable (eds.). Sunderland, Mass.: Sinauer, pp. 515–543.

Hillis, D. M., C. Moritz, and B. K. Mable (eds.). 1996a. *Molecular Systematics,* 2nd Ed. Sunderland, Mass.: Sinauer.

Hooker, J. D. 1853. Introductory essay. In *The Botany of the Antarctic Voyage of H. M. Discovery Ships Erebus and Terror in the Years 1853–55. II Flora Nova Zelandiae.* London: Reeve, pp. 1–34.

Hovenkamp, P. 1997. Vicariance events, not areas, should be used in biogeographical analysis. *Cladistics* 13:67–79.

Huelsenbeck, J. P., B. Larget, and D. Swofford. 2000b. A compound Poisson process for relaxing the molecular clock. *Genetics* 154:1879–1892.

Huelsenbeck, J. P. and R. Nielsen. 1999. Variation in the pattern of nucleotide substitution across sites. *J. Mol. Evol.* 48:86–93.

Huelsenbeck, J. P., B. Rannala, and B. Larget. 2000a. A Bayesian framework for the analysis of cospeciation. *Evolution* 54:352–364.

Huelsenbeck, J. P., F. Ronquist, R. Nielsen, and J. P. Bollback. 2001. Bayesian inference of phylogeny and its impact on evolutionary biology. *Science* 294:2310–2314.

Humphries, C. J. 1981. Biogeographical methods and the southern beeches. In *Advances in Cladistics, Proceedings of the First Meeting of the Hennig Society,* V. A. Funk and D. R. Brooks (eds.). Bronx: New York Botanical Garden, pp. 177–207.

——— 1983. Biogeographical explanations and the southern beeches. In *Evolution, Time and Space: The Emergence of the Biosphere,* R. W. Sims, J. H. Price, and P. E. S. Whalley (eds.). New York: Academic Press, Systematics Association Special Volume n. 23, pp. 335–365.

——— 1989. Any advance on assumption 2? *J. Biogeog.* 16:101–102.

——— 2000. Form, space and time; which comes first? *J. Biogeog.* 27:11–15.

Humphries, C. J., P. Y. Ladiges, M. Roos, and M. Zandee. 1988. Cladistic biogeography. In *Analytical Biogeography: An Integrated Approach to the Study of Animal and Plant Distributions,* A. A. Myers and P. S. Giller (eds.). New York: Chapman & Hall, pp. 371–404.

Humphries, C. J. and L. R. Parenti. 1986. *Cladistic Biogeography.* Oxford: Clarendon Press.

——— 1999. *Cladistic Biogeography,* 2nd ed. New York: Oxford University Press.

Humphries, C. J. and O. Seberg. 1989. Graphs and generalized tracks: Some comments on method. *Syst. Zool.* 38:69–76.

Humphries, C. J., P. H. Williams, and R. I. Vane-Wright. 1995. Measuring biodiversity value for conservation. *Annu. Rev. Ecol. Syst.* 26:93–111.

Hunn, C. A. and P. Upchurch. 2001. The importance of time/space in diagnosing the causality of phylogenetic events: Towards a "chronobiogeographical" paradigm? *Syst. Biol.* 50: 391–407.

Jordan, G. J. and R. S. Hill. 1999. The phylogenetic affinities of *Nothofagus* (Nothofagaceae) leaf fossils based on combined molecular and morphological data. *International J. Plant Sci.* 160:1177–1188.

Juan, C., P. Oromi, and G. M. Hewitt. 1995. Mitochondrial DNA phylogeny and sequencial colonization of Canary Islands by darking beetles of the genus *Pimelia* (Tenebrionidae). *Proc. R. Soc. London* B 261:173–180.

Katinas, L. and J. V. Crisci. 1999. La filogenia frente a la justicia: ¿El fin del caso? *Ciencia Hoy* 9:28–33.

Katinas, L. and J. V. Crisci. 2000. Cladistic and biogeographic analyses of the genera *Moscharia* and *Polyachyrus* (Asteraceae, Mutisieae). *Syst. Bot.* 25:33–46.

Katinas, L., J. J. Morrone, and J. V. Crisci. 1999. Track analysis reveals the composite nature of the Andean biota. *Aust. J. Bot.* 47:111–130.

Keast, A. 1991. Panbiogeography: Then and now. *Quart. Rev. Biol.* 66:467–472.

Kitching, I. J., P. L. Forey, C. J. Humphries, and D. M. Williams. 1998. *Cladistics: The Theory and Practice of Parsimony Analysis,* 2nd Ed. New York: Oxford University Press.

Kluge, A. G. 1993. Three-taxon transformation in phylogenetic inference: Ambiguity and distorsion as regards explanatory power. *Cladistics* 9:246–259.

Kluge, A. G. and J. S. Farris. 1969. Quantitative phyletics and the evolution of anurans. *Syst. Zool.* 18:1–32.

Knox, E. B. and J. D. Palmer. 1998. Chloroplast DNA evidence on the origin and radiation of the giant Lobelias in Eastern Africa. *Syst. Bot.* 23:109–149.

Krzywinski, J., R. C. Wilkerson, and N. J. Besansky. 2001. Toward understanding Anophelinae (Diptera, Culicidae) phylogeny: Insights from nuclear single-copy genes and the weight of evidence. *Syst. Biol.* 50:540–556.

Kuhn, T. S. 1970. *The Structure of Scientific Revolutions,* 2nd Ed. Chicago: The University of Chicago Press.

Ladiges, P. Y., G. Nelson, and J. Grimes. 1997. Subtree analysis, *Nothofagus* and Pacific biogeography. *Cladistics* 13:125–129.

Ladiges, P. Y., S. M. Prober, and G. Nelson. 1992. Cladistic and biogeographic analyses of the "blue ash" Eucalypts. *Cladistics* 8:103–124.

Lakatos, I. 1970. Falsification and the methodology of scientific research programs. In *Criticism and the Growth of Knowledge,* I. Lakatos and A. Musgrave (eds.). Cambridge: Cambridge University Press, pp. 91–176.

Larget, B. and D. L. Simon. 1999. Markov Chain Monte Carlo algorithms for the Bayesian analysis of phylogenetic trees. *Mol. Biol. Evol.* 16:750–759.

Lavin, M., M. Thulin, J. N. Labat, and R. T. Pennington. 2000. Africa, the odd man out: Molecular biogeography of Dalbergioid legumes (Fabaceae) suggests otherwise. *Syst. Bot.* 25:449–467.

Lewis, P. O. 2001a. A likelihood approach to estimating phylogeny from discrete morphological character data. *Syst. Biol.* 50:913–925.

——— 2001b. Phylogenetic systematics turns over a new leaf. *Trends Ecol. Evol.* 16:30–37.

Lieberman, B. S. 2000. *Paleobiogeography: Using Fossils to Study Global Change, Plate Tectonics, and Evolution.* New York: Kluwer Academic Press.

Lieberman, B. S. and N. Eldredge. 1996. Trilobite biogeography in the Middle Devonian: Geological processes and analytical methods. *Paleobiology* 22:66–79.

Liebherr, J. K. 1991. A general area cladogram for Montane Mexico based on distributions in the platynine genera *Elliptoleus* and *Calathus* (Coleoptera: Carabidae). *Proc. Entomol. Soc. Wash.* 93:390–406.

——— 1994a. Biogeographic patterns of Montane Mexican and Central American Carabidae (Coleoptera). *The Canadian Entomologist* 126:841–860.

——— 1994b. Identification of New World *Agonum,* review of the Mexican fauna, and description of *Incagonum,* new genus, from South America (Coleoptera: Carabidae: Platynini). *J. New York Entomol. Soc.* 102:1–55.

Linder, H. P. 1999. *Rytidosperma vickeryae* – a new Danthonioid grass from Kosciusko (New South Wales, Australia): Morphology, phylogeny and biogeography. *Aust. Syst. Bot.* 12:743–755.

Linder, H. P. 2001. On areas of endemism, with an example from the African Restionaceae. *Syst. Biol.* 50:892–912.

Linder, H. P. and M. C. Crisp. 1995. *Nothofagus* and Pacific biogeography. *Cladistics* 11:5–32.

Linder, H. P. and D. M. Mann. 1998. The phylogeny and biogeography of *Thamnochortus* (Restionaceae). *Biol. J. Linnean Soc.* 128:319–357.

Luna, I. and O. Alcántara. 2001. Análisis de simplicidad de endemismos (PAE) para establecer un modelo de vicarianza preliminar del bosque mesófilo de montaña mexicano. In *Introducción a la Biogeografía en América Latina: Teorías, Conceptos, Métodos y Aplicaciones,* J. Llorente-Bousquets and J. J. Morrone (eds.). México: Las Prensas de Ciencias, pp. 273–277.

Lundberg, J. G. 1972. Wagner networks and ancestors. *Syst. Zool.* 21:398–413.

MacArthur, R. H. and E. O. Wilson. 1967. *The Theory of Island Biogeography.* Princeton: Princeton University Press.

MacRae, A. F. and W. W. Anderson. 1988. Evidence for non-neutrality of mitochondrial DNA haplotypes in *Drosophila pseudoobscura. Genetics* 120:485–494.

Malhotra, A. and R. S. Thorpe. 1994. Parallels between island lizards suggests selection on mitochondrial DNA and morphology. *Proc. R. Soc. Lond.* B 257:37–42.

Mallet, J. 1995. A species definition for a modern synthesis. *Trends Ecol. Evol.* 10:294–299.

——— 1996. The genetics of biological diversity: From varieties to species. In *A Biology of Numbers and Differences,* K. J. Gaston (ed.). Oxford: Blackwell, pp. 13–53.

Manos, P. S. 1997. Systematics of *Nothofagus* (Nothofagaceae) based on rDNA spacer sequences (ITS): Taxonomic congruence with morphology and plastid sequences. *Amer. J. Bot.* 84:1137–1155.

Marino, P. I., G. R. Spinelli, and P. Posadas. 2001. Distributional patterns of species of Ceratopogonidae (Diptera) in southern South America. *Biogeographica* 77:113–122.

Marshall, C. J. and J. K. Liebherr. 2000. Cladistic biogeography of the Mexican transition zone. *J. Biogeog.* 27: 203–216.

Marshall, L. G. and R. L. Cifelli. 1990. Analysis of changing diversity patterns in Cenozoic Land Mammal Age faunas, South America. *Palaeovertebrata* 19:169–210.

Marshall, L. G. and T. Sempere. 1993. Evolution of the Neotropical Cenozoic Land Mammal Fauna in its geochronologic, stratigraphic, and tectonic context. In *Biological Relationships Between Africa and South America,* P. Goldblatt (ed.). New Haven: Yale University Press, pp. 329–392.

Martin, P. G. and J. M. Dowd. 1993. Using sequences of *rbc*L to study phylogeny and biogeography of *Nothofagus* species. *Aust. Syst. Bot.* 6:441–447.

Matthew, W. D. 1915. Climate and evolution. *Ann. New York Acad. Sci.* 24:171–318.

Mau, B. and M. Newton. 1997. Phylogenetic inference for binary data on dendrograms using Markov chain Monte Carlo. *J. Comp. Graph. Stat.* 6:122–131.

Mayden, R. L. 1991. The wilderness of panbiogeography: A synthesis of space, time and form. *Syst. Zool.* 40:503–519.

Mayr, E. 1946. History of the North American bird fauna. *Wilson Bull.* 58:3–41.

McAllister, D. E., S. P. Platania, F. W. Schueler, M. E. Baldwin, and D. S. Lee. 1986. Ichthyofaunal patterns on a geographic grid. In *The Zoogeography of North American Freshwater Fishes,* C. H. Hocutt and E. O. Willey (eds.). New York: John Willey and Sons, pp. 17–51.

McLaughlin, S. P. 1992. Are floristic areas hierarchically arranged? *J. Biogeog.* 19:21–32.

Meacham, C. 1984. Evaluating characters by character compatibility analysis. In *Cladistics: Perspectives on the Reconstruction of Evolutionary History,* T. Duncan and T. F. Stuessy (eds.). New York: Columbia University Press, pp. 152–165.

Melville, R. 1973. Continental drift and plant distribution. In *Implications of Continental Drift to the Earth Sciences,* D. H. Tarling and S. K. Runcorn (eds.). London and New York: Academic Press, vol. 1, NATO Advanced Study Institutes Series, pp. 439–446.

Mickevich, M. F. 1981. Quantitative phylogenetic biogeography. In *Advances in Cladistics: Proceedings of the First Meeting of the Willi Hennig Society,* V. A. Funk and D. R. Brooks (eds.). Bronx, New York: New York Bot. Gard., pp. 202–222.

Miller, R. L. 1994. Setting the scene. In *Mapping the Diversity of Nature,* R. I. Miller (ed.). New York: Chapman & Hall, pp. 1–17.

Miranda-Esquivel, D. R. 1999. *Análisis Filogenético de la Tribu Simuliini (sensu Crosskey, 1987) para las Regiones Zoogeográficas Neotropical, Etiópica y Australiana.* La Plata, Argentina: Ph. D. thesis disertation, Universidad Nacional de La Plata, Argentina.

Moore, D. M. 1972. Connection between cool temperate floras, with particular reference to southern South America. In *Taxonomy, Phytogeography and Evolution,* D. H. Valentine (ed.). London and New York: Academic Press, pp. 115–138.

Moore, D. M. 1983. *Flora of Tierra del Fuego*. England and USA: Anthony Nelson, Missouri Botanical Garden.

Morell, P. L., J. M. Porter, and E. A. Friar. 2000. Intercontinental dispersal: The origin of the widespread South American plant species *Gilia laciniata* (Polemoniaceae) from a rare California and Oregon coastal endemic. *Plant Syst. Evol.* 224:13–32.

Moritz, C. 1995. Uses of molecular phylogenies for conservation. *Philos. Trans. R. Soc. London Ser. B* 349:113–118.

Morrone, J. J. 1992. Revisión sistemática, análisis cladístico y biogeografía histórica de los géneros *Falklandius* Enderlein y *Lanteriella gen. nov.* (Coleoptera; Curculionidae). *Acta Ent. Chilena* 17:157–174.

——— 1993a. Revisión sistemática de un nuevo género de Rhytirrhinini (Coleoptera: Curculionidae), con un análisis biogeográfico del dominio Subantártico. *Bol. Soc. Biol. Concepción* 64:121–145.

——— 1993b. Cladistic and biogeographic analyses of the weevil genus *Listroderes* Schoenherr (Coleoptera: Curculionidae). *Cladistics* 9:397–411.

——— 1993c. Beyond binary oppositions. *Cladistics* 9:437–438.

——— 1994a. On the identification of areas of endemism. *Syst. Biol.* 43:438–441.

——— 1994b. Distributional patterns of species of Rhytirrhinini (Coleoptera: Curculionidae) and the historical relationships of the Andean provinces. *Global Ecol. Biogeogr. Lett.* 4:188–194.

——— 1996a. The biogeographical Andean subregion: A proposal exemplified by Arthropod taxa (Arachnida, Crustacea, and Hexapoda). *Neotropica* 42:103–114.

——— 1996b. Austral biogeography and relict weevil taxa (Coleoptera: Nemonychidae, Belidae, Brentidae, and Caridae). *J. Comp. Biol.* 1:123–128.

——— 1998. On Udvardy´s Insulantarctica province: A test from the weevils (Coleoptera: Curculionoidea). *J. Biogeog.* 25:947–955.

——— 2000. La importancia de los atlas biogeográficos para la conservación de la biodiversidad. In *Hacia un Proyecto CYTED para el Inventario y Estimación de la Diversidad Entomológica en Iberoamérica: PrIBES 2000,* F. M. Piera, J. J. Morrone, and A. Melic (eds.), vol. 1. Zaragoza, España: S.E.A., Monografías Tercer Milenio, pp. 69–78.

Morrone, J. J. and J. M. Carpenter. 1994. In search of a method for cladistic biogeography: An empirical comparison of component analysis, Brooks parsimony analysis, and three-area statements. *Cladistics* 10:99–153.

Morrone, J. J., M. M. Cigliano, and J. V. Crisci. 1992. Cladismo y diversidad biológica. *Ciencia Hoy* 4:26–34.

Morrone, J. J. and M. del C. Coscarón. 1996. Distributional patterns of the American Peiratinae (Heteroptera: Reduviidae). *Zool. Med. Leiden* 70:1–15.

Morrone, J. J. and J. V. Crisci. 1990. Panbiogeografía fundamentos y métodos. *Evolución Biológica* (Bogotá) 4:119–140.

——— 1995. Historical biogeography: Introduction to methods. *Annu. Rev. Ecol. Syst.* 26:373–401.

Morrone, J. J., D. Espinosa-Organista, C. Aguilar-Zuñiga, and J. Llorente-Bousquets. 1999. Preliminary classification of the Mexican biogeographic provinces: A parsimony analysis of endemicity based on plant, insect and bird taxa. *Southw. Naturalist* 44:507–544.

Morrone, J. J., D. Espinosa-Organista, and J. Llorente-Bousquets. 1996. *Manual de Biogeografía Histórica,* 1st Ed. México: Universidad Nacional Autónoma de México.

Morrone, J. J., L. Katinas, and J. V. Crisci. 1997. A cladistic biogeographic analysis of Central Chile. *J. Comp. Biol.* 2:25–41.

Morrone, J. J. and E. C. Lopretto. 1995. Parsimony analysis of endemicity of freshwater Decapoda (Crustacea: Malacostraca) from southern South America. *Neotropica* 41:3–8.

Morrone, J. J. and J. Márquez. 2001. Halffters´s Mexican Transition Zone, beetle generalized tracks, and geographical homology. *J. Biogeog.* 28:635–650.

Morrone, J. J., S. Roig-Juñent, and J. V. Crisci. 1994. Cladistic biogeography of terrestrial Subantarctic beetles (Insecta: Coleoptera) from South America. *Ntl. Geog. Research and Exploration* 10:104–115.

Morrone, J. J. and E. Urtubey. 1997. Historical biogeography of the northern Andes: A cladistic analysis based on five genera of Rhytirrhinini (Coleoptera: Curculionidae) and *Barnadesia* (Asteraceae). *Biogeographica* 73:115–121.

Müller, F. 1864. Für Darwin. In *Fritz, Müller, Werke, Briefe, und Leben,* A. Moller (ed.). Jena: Gustav Fischer, pp. 202–263.

Müller, P. 1973. *The Dispersal Centres of Terrestrial Vertebrates in the Neotropical Realm: A Study in the Evolution of the Neotropical Biota and its Native Landscapes.* Junk: The Hague.

Muss, A., D. R. Robertson, C. A. Stepien, P. Wirtz, and B. W. Bowen. 2001. Phylogeography of *Ophioblennius:* The role of ocean currents and geography in reef fish evolution. *Evolution* 55:561–572.

Myers, A. A. 1991. How did Hawaii accumulate its biota? A test from the Amphipoda. *Global Ecol. Biogeogr. Lett.* 1:24–29.

Myers, A. A. and P. S. Giller (eds.). 1988. *Analytical Biogeography: An Integrated Ap-*

proach to the Study of Animal and Plant Distributions. New York: Chapman & Hall.

Nelson, G. 1973. Comments on Leon Croizat's biogeography. *Syst. Zool.* 22:312–320.

——— 1974. Historical Biogeography: An alternative formalization. *Syst. Zool.* 23:555–558.

——— 1978. From Candolle to Croizat: Comments on the history of biogeography. *J. Hist. Biol.* 11:269–305.

Nelson, G. and P. Y. Ladiges. 1991a. Three-area statements: Standard assumptions for biogeographic analysis. *Syst. Zool.* 40:470–485.

——— 1991b. Standard assumptions for biogeographic analyses. *Aust. Syst. Bot.* 4:41–58.

——— 1992. *TAS and TAX: MSDos computer programs for cladistics.* New York and Melbourne: Published by the authors.

——— 1995. *TASS (Three area subtrees), version 2.0.* New York and Melbourne: Published by the authors.

——— 1996. Paralogy in cladistic biogeography and analysis of paralogy-free subtrees. *Am. Museum Novit.* 3167:1–58.

——— 2001. Gondwana, vicariance biogeography and the New York School revisited. *Aust. J. Bot.* 49:389–409.

Nelson, G. and N. I. Platnick. 1981. *Systematics and Biogeography: Cladistics and Vicariance.* New York: Columbia University Press.

——— 1984. *Biogeography.* Burlington, North Carolina: Carolina Biology Readers n. 119. Biological Supply Company.

——— 1991. Three-taxon statements: A more precise use of parsimony? *Cladistics* 7:351–366.

Nordlander, G., Z. Liu, and F. Ronquist. 1996. Phylogeny and historical biogeography of the cynipoid wasp family Ibaliidae (Hymenoptera). *Syst. Entomol.* 21:151–166.

Oakeshott, M. J. 1959. *The Voice of Poetry in the Conversation of Mankind.* London: Bowes & Bowes.

Oliver, W. R. B. 1925. Biogeographical relations of the New Zealand region. *Bot. J. Linn. Soc. Lond.* 47:99–139.

Olmstead, R. G. and J. D. Palmer. 1997. Implications for the phylogeny, classification, and biogeography of *Solanum* from cpDNA restriction site variation. *Syst. Bot.* 22:19–29.

Page, R. D. M. 1987. Graphs and generalized tracks: Quantifying Croizat's panbiogeography. *Syst. Zool.* 36:1–17.

——— 1988. Quantitative cladistic biogeography: Constructing and comparing area cladograms. *Syst. Zool.* 37:254–270.

——— 1989. *COMPONENT user's manual, Release 1.5.* Auckland: Published by the author.

——— 1990. Component analysis: A valiant failure? *Cladistics* 6:119–136.

——— 1993. *COMPONENT user's manual, release 2.0.* London: Natural History Museum.

——— 1994a. Maps between trees and cladistic analysis of historical associations among genes, organisms, and areas. *Syst. Biol.* 43:58–77.

——— 1994b. Parallel phylogenies: Reconstructing the history of host-parasite assemblages. *Cladistics* 10:155–173.

——— 2000. Extracting species trees from complex gene trees: Reconciled trees and vertebrate phylogeny. *Mol. Phylogenet. Evol.* 14:89–106.

Page, R. D. M. and M. A. Charleston. 1998. Trees within trees: Phylogeny and historical associations. *Trends Ecol. Evol.* 13:356–359.

Page, R. D. M. and E. C. Holmes. 1998. *Molecular Evolution: A Phylogenetic Approach.* Oxford: Blackwell Science.

Page, R. D. M. and C. Lydeard. 1994. Towards a cladistic biogeography of the Caribbean. *Cladistics* 10:21–41.

Palmer, M. and Y. Cambefort. 2000. Evidence for reticulate palaeogeography: Beetle diversity linked to connection-disjunction cycles of the Gibraltar strait. *J. Biogeog.* 27:403–416.

Pascual, R. and E. Ortiz Jaureguizar. 1990. Evolving climates and mammal faunas in South America. *J. Human Evol.* 19:23–60.

Pascual, R., E. Ortiz Jaureguizar, and J. L. Prado. 1996. Land mammals: Paradigm of Cenozoic South American geobiotic evolution. *Münchner Geowiss. Abh.* (A) 30:265–319.

Patterson, A. M., G. P. Wallis, L. J. Wallis, and R. D. Gray. 2000. Seabird and louse coevolution: Complex histories revealed by 12S rRNA sequences and reconciliation analyses. *Syst. Biol.* 49:383–399.

Patterson, C. 1981. Methods of paleobiogeography. In *Vicariance Biogeography: A Critique,* G. Nelson and D. E. Rosen (eds.). New York: Columbia University Press, pp. 446–489.

Philipson, W. R. and N. N. Philipson. 1988. A classification of the genus *Nothofagus* (Fagaceae). *Bot. J. Linn. Soc.* 98:27–36.

Platnick, N. I. 1991. On areas of endemism. *Aust. Syst. Bot.* 4: without numeration.

——— 1992. Patterns of biodiversity. In *Systematics, Ecology, and the Biodiversity Crisis,* N. Eldredge (ed.). New York: Columbia University Press, pp. 15–25.

Platnick, N. I. and G. Nelson. 1978. A method of analysis for historical biogeography. *Syst. Zool.* 27:1–16.

——— 1984. Composite areas in vicariance biogeography. *Syst. Zool.* 33:328–335.

——— 1988. Spanning tree biogeography: Shortcut, detour or dead-end? *Syst. Zool.* 37:410–419.

Pole, M. 1994. The New Zealand flora-entirely long distance dispersal? *J. Biogeog.* 21:625–635.

Polhemus, D. A. 1996. Island arcs, and their influence on Indo-Pacific biogeography. In *The Origin and Evolution of Pacific Island Biotas, New Guinea to Eastern Polynesia: Patterns and Processes,* A. Keast and S. E. Miller (eds.). Amsterdam: SPB Academic Publishing.

Popper, K. R. 1959. *The Logic of Scientific Discovery.* London: Hutchinson.

Posada, D., K. A. Crandall, and A. R. Templeton. 2000. GeoDis: A program for the Cladistic Nested Analysis of the geographical distribution of genetic haplotypes. *Mol. Ecol.* 9:487–488.

Posadas, P. 1996. Distributional patterns of vascular plants in Tierra del Fuego: A study applying parsimony analysis of endemicity (PAE). *Biogeographica* 72:161–177.

Posadas, P., J. M. Estévez, and J. J. Morrone. 1997. Distributional patterns and endemism areas of vascular plants in the Andean subregion. *Fontqueria* 48:1–10.

Posadas, P. and D. R. Miranda-Esquivel. 1999. El PAE (Parsimony Analysis of Endemicity) como una herramienta en la evaluación de la biodiversidad. *Rev. Chilena Hist. Nat.* 72:483–490.

Posadas, P., D. R. Miranda-Esquivel, and J. V. Crisci. 2001. Using phylogenetic diversity measures to set priorities in conservation: An example from southern South America. *Conserv. Biol.* 15:1325–1334.

Posadas, P. and J. J. Morrone. 2001. Biogeografía cladística de la subregión Subantártica: Un análisis basado en taxones de la Familia Curculionidae (Insecta: Coleoptera). In *Introducción a la Biogeografía en América Latina: Teorías, Conceptos, Métodos y Aplicaciones,* J. Llorente-Bousquets and J. J. Morrone (eds.). México: Las Prensas de Ciencias, pp. 267–271.

Prance, G. 2000. The failure of biogeographers to convey the conservation message. *J. Biogeog.* 27:51–57.

Rahel, F. J. 2000. Homogenization of fish faunas across the United States. *Science* 288:854–856.

Rannala, B. and Z. Yang. 1996. Probability distribution of molecular evolutionary trees: A new method for phylogenetic inference. *J. Mol. Evol.* 43:304–311.

Rapoport, E. H. 1975. *Areografía: Estrategias Geográficas de las Especies.* México: Fondo de Cultura Económica.

Rapoport, E. H. and J. A. Monjeau. 2001. Areografía. In *Introducción a la Biogeografía en América Latina: Teorías, Conceptos, Métodos y Aplicaciones,* J. Llorente-Bousquets and J. J. Morrone (eds.). México: Las Prensas de Ciencias, pp. 23–30.

Raven, P. H. and D. I. Axelrod. 1972. Plate tectonics and Australasian paleogeography: The complex biogeographic relations of the region reflect its geologic history. *Science* 176:1379–1386.

——— 1974. Angiosperm biogeography and past continental movements. *Ann. Missouri Bot. Gard.* 61:539–673.

Renner, S. S., G. Clausing, and K. Meyer. 2001. Historical biogeography of Melastomataceae: The roles of Tertiary migration and long-distance dispersal. *Amer. J. Bot.* 88:1290–1300.

Repetur, C. P., P. C. van Welzen, and E. F. de Vogel. 1997. Phylogeny and historical biogeography of the genus *Bromheadia* (Orchidaceae). *Syst. Bot.* 22:465–477.

Rich, S. M., D. A. Caporale, S. R. Telford III, T. D. Kocher, D. L. Hart, and A. Spielman. 1995. Distribution of *Ixodes ricinus*-like ticks of eastern North America. *Proc. Natl. Acad. Sci. USA* 92:6284–6288.

Riddle, B. R. and R. L. Honeycutt. 1990. Historical biogeography in North American arid regions: An approach using mitochondrial-DNA phylogeny in grasshopper mice (genus *Onychomys*). *Evolution* 44:1–15.

Ringuelet, R. A. 1956. Panorama zoogeográfico de la provincia de Buenos Aires. *Notas Mus. La Plata* 18:1–15.

Roig-Juñent, S. 1999. Areas de distribución y áreas de endemismo: Definiciones y criterios para su reconocimiento. Libro de resúmenes, *II Reunión Argentina de Cladística y Biogeografía,* Buenos Aires, Argentina, pp. 29.

Roig-Juñent, S., J. V. Crisci, P. Posadas, and S. Lagos. 2002. Areas de distribución y áreas de endemismo en zonas continentales. In *Proyecto de Red Iberoamericana de Biogeografía y Entomología Sistemática, Vol. 2,* S. A. Vanin, J. M. Lobo, and A. Melic (eds.). Spain: Sociedad Entomológica Aragonesa y CYTED, pp. 247–266.

Roig-Juñent, S., G. Flores, S. Claver, G. Debandi, and A. Marvaldi. 2000. Monte Desert (Argentina): Insect biodiversity and natural areas. *J. Arid Environm.* 46:1–18.

Romesburg, H. C. 1984. *Cluster Analysis for Researchers.* Belmont: Lifetime Learning Publications.

Ron, S. R. 2000. Biogeographic area relationships of lowland Neotropical rainforest based on raw distributions of vertebrate groups. *Biol. J. Linnean Soc.* 71:379–402.

Ronquist, F. 1994. Ancestral areas and parsimony. *Syst. Biol.* 43:267–274.

——— 1995. Ancestral areas revisited. *Syst. Biol.* 44:572–575.

——— 1996. *DIVA, version 1.1.* Computer program and manual available by anonymous FTP from Uppsala University (ftp.systbot.uu.se).

——— 1997a. Phylogenetic approaches in coevolution and biogeography. *Zoologica Scripta* 26:313–322.

——— 1997b. Dispersal-vicariance analysis: A new approach to the quantification of historical biogeography. *Syst. Biol.* 46:195–203.

——— 1998. Three-dimensional cost-matrix optimization and maximum cospeciation. *Cladistics* 14:167–172.

Ronquist, F. and S. Nylin. 1990. Process and pattern in the evolution of species association. *Syst. Zool.* 39:323–344.

Rosen, B. R. 1988. From fossils to earth history: Applied historical biogeography. In *Analytical Biogeography: An Integrated Approach to the Study of Animal and Plant Distributions,* A. A. Myers and P. S. Giller (eds.). New York: Chapman & Hall, pp. 437–481.

Rosen, B. R. and A. B. Smith. 1988. Tectonics from fossils? Analysis of reef-coral and sea-urchin distributions from Late Cretaceous to Recent, using a new method. In *Gondwana and Tethys,* M. G. Audley-Charles and A. Hallam (eds.). Oxford: Oxford University Press, pp. 275–306.

Rosen, D. E. 1976. A vicariance model of Caribbean biogeography. *Syst. Zool.* 24:431–464.

——— 1978. Vicariant patterns and historical explanation in biogeography. *Syst. Zool.* 27:159–188.

Ross, H. H. 1974. *Biological Systematics.* Mass.: Addison-Wesley Reading.

Rozas, A., J. M. Hernández, V. M. Cabrera, and A. Prerosti. 1990. Colonization in America by *Drosophila subobscura:* Effect of the founder event on mitochondrial DNA polymorphism. *Mol. Biol. Evol.* 7:103–109.

Ryder, O. A. 1986. Species conservation and the dilemma of subspecies. *Trends Ecol. Evol.* 1:9–10.

Sala, O. E., F. S. Chapin III, J. J. Armesto, E. Berlow, J. Bloomfield, R. Dirzo, E. Huber-Sanwald, L. F. Huenneke, R. B. Jackson, A. Kinzig, R. Leemans, D. M. Lodge, H. A. Mooney, M. Oesterheld, N. LeRoy Poff, M. T. Sykes, B. H. Walker, M. Walker, and D. H. Wall. 2000. Global biodiversity scenarios for the year 2100. *Science* 287:1770–1774.

Salisbury, B. A. 1999. *SECANT: Strongest Evidence Compatibility Analytic Tool. Version 2.2.* New Haven: Department of Ecology and Evolutionary Biology, Yale University.

Sanderson, M. J. 1998. Estimating rate and time in molecular phylogenies: Beyond the molecular clock? In *Molecular Systematics of Plants II: DNA sequencing,* D. E. Soltis, P. S. Soltis, and J. J. Doyle (eds.). Boston, Dordrecht, London: Kluwer Academic Publishers, pp. 242–264.

Sanmartín, I., H. Enghoff, and F. Ronquist. 2001. Patterns of animal dispersal, vicariance and diversification in the Holarctic. *Biol. J. Linnean Soc.* 73:345–390.

Schuh, R. T. 2000. *Biological Systematics: Principles and Applications.* Ithaca: Cornell University Press.

Schuster, R. M. 1976. Plate tectonics and its bearing on the geographical origin and dispersal of angiosperms. In *Origin and Early Evolution of Angiosperms,* C. B. Beck (ed.). New York: Columbia University Press, pp. 48–138.

Seberg, O. 1986. A critique of the theory and methods of panbiogeography. *Syst. Zool.* 35:369–380.

——— 1991. Biogeographic congruence in the South Pacific. *Aust. Syst. Bot.* 4:127–136.

Sequeira, A. S. and B. D. Farrel. 2001. Evolutionary origins of Gondwanan interactions: How old are *Araucaria* beetle herbivores? *Biol. J. Linn. Soc.* 74:459–474.

Setoguchi, H., M. Ono, Y. Doi, H. Koyama, and M. Tsuda. 1997. Molecular phylogeny of *Nothofagus* based on the *atpB-rbcL* intergenic spacer of the chloroplast DNA. *J. Plant Res.* 110:469–484.

Sharp, P. M., D. C. Shields, K. H. Wolfe, and W. H. Li. 1989. Chromosomal location and evolutionary rate variation in enterobacterial genes. *Science* 258:808–810.

da Silva, M. N. F. and L. Patton. 1993. Amazonian phylogeography: mtDNA sequence variation in arboreal Echimiyd rodents (Caviomorpha). *Mol. Biol. Evol.* 2:243–255.

——— 1998. Molecular phylogeography and the evolution and conservation of Amazonian mammals. *Mol. Ecol.* 7:475–486.

Simpson, B. B. 1975. Pleistocene changes in the flora of high tropical Andes. *Paleobiology* 1:273–294.

Simpson, G. G. 1944. *Tempo and Mode in Evolution.* New York: Columbia University Press.

——— 1964. *Evolución y Geografía.* Buenos Aires, Argentina: EUDEBA.

——— 1965. *The Geography of Evolution.* Philadelphia and New York: Chilton.

Simpson Vuilleumier, B. B. 1971. Pleistocene changes in the fauna and flora of South America. *Science* 173:771–774.

Smith, A. B. 1992. Echinoid distribution in the Cenomanian: An analytical study in biogeography. *Palaeogeog. Palaeoclimatol. Palaeoecol.* 92:263–276.

Soulé, M. E. 1991. Conservation: Tactics for a constant crisis. *Science* 253:744–750.

Spellerberg, I. F. and J. W. D. Sawyer. 1999. *An Introduction to Applied Biogeography.* Cambridge: Cambridge University Press.

Stheli, F. G. and S. D. Webb (eds.). 1985. *The Great American Biotic Interchange.* New York and London: Plenum Press.

Stuessy, T. F., T. Sang, and M. L. De Vore. 1996. Phylogeny and biogeography of the subfamily Barnadesioideae with implications for early evolution of the Compositae. In *Compositae: Systematics. Proceedings of the International Compositae Conference Kew, 1994,* D. J. N. Hind and H. J. Beentje (eds.). Royal Botanical Gardens Kew, Vol. 1, pp. 463–490.

Sun, H., W. McLewin, and M. F. Fay. 2001. Molecular phylogeny of *Helleborus* (Ranunculaceae), with an emphasis on the East Asian–Mediterranean disjunction. *Taxon* 50:1001–1018.

Swenson, U., A. Backlund, S. McLoughlin, and R. S. Hill. 2001. *Nothofagus* biogeography revisited with special emphasis on the enigmatic distribution of subgenus *Brassospora* in New Caledonia. *Cladistics* 17:28–47.

Swenson, U. and K. Bremer. 1998. Pacific biogeography of the Asteraceae genus *Abrotanella* (Senecioneae, Blemnospermatinae). *Syst. Bot.* 22:493–508.

Swenson, U. and R. S. Hill. 2001. Most parsimonious areagrams versus fossils: The case of *Nothofagus* (Nothofagaceae). *Aust. J. Bot.* 49:367–376.

Swenson, U., R. S. Hill, and S. McLoughlin. 2000. Ancestral area analysis of *Nothofagus* (Nothofagaceae). *Aust. Syst. Bot.* 13:469–478.

Swofford, D. L. 2000. *PAUP* (Phylogenetic Analysis Using Parsimony and other methods).* Sunderland, Mass.: Sinauer.

Swofford, D. L., G. J. Olsen, P. J. Waddell, and D. M. Hillis. 1996. Phylogenetic inference. In *Molecular Systematics,* 2nd Ed., D. M. Hillis, C. Moritz, and B. K. Mable (eds.). Sunderland, Mass.: Sinauer, pp. 407–514.

Swofford, D. L., P. J. Waddel, J. P. Huelsenbeck, P. G. Foster, P. O. Lewis, and J. S. Rogers. 2001. Bias in phylogenetic estimation and its relevance to the choice between parsimony and likelihood methods. *Syst. Biol.* 50:525–539.

Takezaki, N., A. Rzhetsky, and M. Nei. 1995. Phylogenetic test of the molecular clock and linearized trees. *Molec. Biol. Evol.* 12:823–833 (the computer program Lintre is available from http://www.cib.nig.ac.jp/dda/ntakezak/ntakezak.html).

Takhtajan, A. 1986. *Floristic Regions of the World.* Berkeley: University California Press.

Templeton, A. R. 1998. Nested clade analysis of phylogeographic data: Testing hypotheses about gene flow and population history. *Mol. Ecol.* 7:381–398.

——— 2001. Using phylogeographic analyses of gene trees to test species status and processes. *Mol. Ecol.* 10:779–791.

Templeton, A. R., E. Routman, and C. A. Phillips. 1995. Separating population structure from population history: A cladistic analysis of the geographical distribution of mitochondrial DNA haplotypes in the Tiger salamander, *Ambystoma tigrinum. Genetics* 140:767–782.

Trejo-Torres, J. C. and J. D. Ackerman. 2001. Biogeography of the Antillas based on a parsimony analysis of orchid ditributions. *J. Biogeog.* 28:775–794.

Tuffley, C. and M. A. Steel. 1997. Links between maximum likelihood and maximum parsimony under a simple model of site substitution. *Bull. Math. Biol.* 59:581–607.

Udvardy, M. D. F. 1969. *Dynamic Zoogeography with Special Reference to Land Animals.* New York: Van Nostrand Reinhold.

Ulfstrand, S. 1992. Biodiversity – How to reduce its decline. *Oikos* 63:3–5.

Vander Zanden, M. J., J. M. Casselman, and J. B. Rasmussen. 1999. Stable isotope for the food web consequences of species invasions in lakes. *Nature* 401:464–467.

Vane-Wright, R. I., C. J. Humphries, and P. H. Williams. 1991. What to protect? – Systematics and the agony of choice. *Biol. Conserv.* 55:235–254.

Van Steenis, C. G. G. J. 1962. The land-bridge theory in botany with particular reference to tropical plants. *Blumea* 11:235–542.

Van Steenis, C. G. G. J. 1971. *Nothofagus,* key genus of plant geography, in time and space, living and fossil, ecology and phylogeny. *Blumea* 19:65–98.

van Veller, M. G. P. and D. R. Brooks. 2001. When simplicity is not parsimonious: *A priori* and *a posteriori* methods in historical biogeography. *J. Biogeog.* 28:1–11.

van Veller, M. G. P., D. J. Kornet, and M. Zandee. 2000. Methods in vicariance biogeography: Assessment of the implementation of assumptions 0, 1, and 2. *Cladistics* 16:319–345.

van Veller, M. G. P., M. Zandee, and D. J. Kornet. 1999. Two requirements for obtaining valid common patterns under different assumptions in vicariance biogeography. *Cladistics* 15:393–406.

Vinnersten, A. and K. Bremer. 2001. Age and biogeography of major clades in Liliales. *Amer. J. Bot.* 88:1695–1703.

Vitousek, P., H. A. Mooney, J. Lubchenco, and J. M. Melillo. 1997. Human domination of Earth's ecosystems. *Science* 277:494–499.

Voelker, G. 1999a. Dispersal, vicariance, and clocks: Historical biogeography and speciation in a cosmopolitan Passerine genus (*Anthus:* Motacillidae). *Evolution* 53:1536–1552.

——— 1999b. Molecular evolutionary relationships in the avian genus *Anthus* (Pipits: Motacillidae). *Mol. Phylogenet. Evol.* 11:84–94.

Walker, D. E. and J. C. Avise. 1998. Principles of phylogeography as illustrated by freshwater and terrestrial turtles in the southeastern United States. *Annu. Rev. Ecol. Syst.* 29:23–58.

Wallace, A. R. 1876. *The Geographical Distribution of Animals.* New York: Hafner.

——— 1892. *Island Life.* London: Macmillan.

Waters, J. M., J. A. López, and G. P. Wallis. 2000. Molecular phylogenetics and biogeography of Galaxiid fishes (Osteichtyes: Galaxiidae): Dispersal, vicariance, and the position of *Lepidogalaxias salamandroides. Syst. Biol.* 49:777–795.

Wegener, A. 1915. *Die Entstehung der Kontinente und Ozeane.* Braunschweig: Vieweg & Sohn.

Weston, P. and M. D. Crisp. 1994. Cladistic biogeography of the Waraths (Proteaceae: Embothrieae) and their allies accross the Pacific. *Aust. Syst. Bot.* 7:225–249.

——— 1996. Trans-Pacific biogeographic patterns in the Proteaceae. In *The Origin and Evolution of Pacific Islands Biotas, New Guinea to Polinesia: Patterns and Processes,* A. Keast and S. E. Miller (eds.). Amsterdam: Academic Publishing, pp. 215–232.

Wiley, E. O. 1980. Phylogenetic systematics and vicariance biogeography. *Syst. Bot.* 5:194–220.

——— 1981. *Phylogenetics: The Theory and Practice of Phylogenetic Systematics.* New York: Wiley-Intersci.

——— 1987. Methods in vicariance biogeography. In *Systematics and Evolution: A Matter of Diversity,* P. Hovenkamp (ed.). Utrecht: Institute of Systematic Botany, Utrecht University, pp. 283–306.

——— 1988. Vicariance biogeography. *Annu. Rev. Ecol. Syst.* 19:513–542.

Williams, W. T. and M. B. Dale. 1965. Fundamental problems in numerical taxonomy. *Adv. Bot. Res.* 2:35–68.

Wilson, R. J. 1983. *Introducción a la Teoría de Grafos.* Madrid: Alianza Editorial.

Wolfe, K. H. 1991. Mammalian DNA replication: Mutation biases and the mutation rate. *J. Theor. Biol.* 149:441–451.

Xiang, Q. Y., S. J. Brunsfeld, D. E. Soltis, and P. S. Soltis. 1996. Phylogenetic relationships in *Cornus* based on chloroplast DNA restriction sites: Implications for biogeography and character evolution. *Syst. Bot.* 21:515–534.

Xiang, Q. Y., D. J. Crawford, A. D. Wolfe, Y. C. Tang, and C. W. DePamphilis. 1998. Origin and biogeography of *Aesculus* L. (Hippocastanaceae): A molecular phylogenetic perspective. *Evolution* 52:988–997.

Yang, Z. H. and B. Rannala. 1997. Bayesian phylogenetic inference using DNA sequences: A Markov Chain Monte Carlo method. *Mol. Biol. Evol.* 14:717–724.

Zandee, M. and M. C. Roos. 1987. Component-compatibility in historical biogeography. *Cladistics* 3:305–332.

Zink, R. M. 1996. Comparative phylogeography in North American birds. *Evolution* 50:308–317.

Zink, R. M., R. C. Blackwell-Rago, and F. Ronquist. 2000. The shifting roles of dispersal and vicariance in biogeography. *Proc. R. Soc. Lond.* B 267:497–503.

Zuckerkandl, E. and L. Pauling. 1962. Molecular disease, evolution, and genetic heterogeneity. In *Horizons in Biochemistry,* M. Kasha and B. Pullman (eds.). London and New York: Academic Press, pp. 189–225.

——— 1965. Evolutionary divergence and convergence. In *Evolving Genes and Proteins,* V. Bryson and H. J. Vogel (eds.). London and New York: Academic Press, pp. 97–166.

Zunino, M. 2000. El concepto de área de distribución: Algunas reflexiones teóricas. In *Hacia un Proyecto CYTED para el Inventario y Estimación de la Diversidad Entomológica en Iberoamérica: PrIBES 2000,* F. M. Piera, J. J. Morrone, and A. Melic (eds.), vol. 1. Zaragoza: SEA, Monografías Tercer Milenio, pp. 80–85.

Zunino, M. and A. Zullini. 1995. *Biogeografía: La Dimensione Spaziale dell'Evoluzione.* Milan: Casa Editrice Ambrosiana.

INDEX

Page numbers in bold indicate figures or tables